Innovations in ART Neural Networks

Studies in Fuzziness and Soft Computing

Editor-in-chief
Prof. Janusz Kacprzyk
Systems Research Institute
Polish Academy of Sciences
ul. Newelska 6
01-447 Warsaw, Poland
E-mail: kacprzyk@ibspan.waw.pl

Vol. 3. A. Geyer-Schulz
Fuzzy Rule-Based Expert Systems and Genetic Machine Learning, 2nd ed. 1996
ISBN 3-7908-0964-0

Vol. 4. T. Onisawa and J. Kacprzyk (Eds.)
Reliability and Safety Analyses under Fuzziness, 1995
ISBN 3-7908-0837-7

Vol. 5. P. Bosc and J. Kacprzyk (Eds.)
Fuzziness in Database Management Systems, 1995
ISBN 3-7908-0858-X

Vol. 6. E. S. Lee and Q. Zhu
Fuzzy and Evidence Reasoning, 1995
ISBN 3-7908-0880-6

Vol. 7. B. A. Juliano and W. Bandler
Tracing Chains-of-Thought, 1996
ISBN 3-7908-0922-5

Vol. 8. F. Herrera and J. L. Verdegay (Eds.)
Genetic Algorithms and Soft Computing, 1996
ISBN 3-7908-0956-X

Vol. 9. M. Sato et al.
Fuzzy Clustering Models and Applications, 1997,
ISBN 3-7908-1026-6

Vol. 10. L. C. Jain (Ed.)
Soft Computing Techniques in Knowledge-based Intelligent Engineering Systems, 1997
ISBN 3-7908-1035-5

Vol. 11. W. Mielczarski (Ed.)
Fuzzy Logic Techniques in Power Systems, 1998
ISBN 3-7908-1044-4

Vol. 12. B. Bouchon-Meunier (Ed.)
Aggregation and Fusion of Imperfect Information, 1998
ISBN 3-7908-1048-7

Vol. 13. E. Orłowska (Ed.)
Incomplete Information: Rough Set Analysis, 1998
ISBN 3-7908-1049-5

Vol. 14. E. Hisdal
Logical Structures for Representation of Knowledge and Uncertainty, 1998
ISBN 3-7908-1056-8

Vol. 15. G. J. Klir and M. J. Wierman
Uncertainty-Based Information, 2nd ed. 1999
ISBN 3-7908-1242-0

Vol. 16. D. Driankov and R. Palm (Eds.)
Advances in Fuzzy Control, 1998
ISBN 3-7908-1090-8

Vol. 17. L. Reznik, V. Dimitrov and J. Kacprzyk (Eds.)
Fuzzy Systems Design, 1998
ISBN 3-7908-1118-1

Vol. 18. L. Polkowski and A. Skowron (Eds.)
Rough Sets in Knowledge Discovery 1, 1998
ISBN 3-7908-1119-X

Vol. 19. L. Polkowski and A. Skowron (Eds.)
Rough Sets in Knowledge Discovery 2, 1998
ISBN 3-7908-1120-3

Vol. 20. J. N. Mordeson and P. S. Nair
Fuzzy Mathematics, 1998
ISBN 3-7908-1121-1

Vol. 21. L. C. Jain and T. Fukuda (Eds.)
Soft Computing for Intelligent Robotic Systems, 1998
ISBN 3-7908-1147-5

Vol. 22. J. Cardoso and H. Camargo (Eds.)
Fuzziness in Petri Nets, 1999
ISBN 3-7908-1158-0

Vol. 23. P. S. Szczepaniak (Ed.)
Computational Intelligence and Applications, 1999
ISBN 3-7908-1161-0

Vol. 24. E. Orłowska (Ed.)
Logic at Work, 1999
ISBN 3-7908-1164-5

continued on page 259

Lakhmi C. Jain
Beatrice Lazzerini
Ugur Halici (Eds.)

Innovations in ART Neural Networks

With 69 Figures
and 30 Tables

Springer-Verlag Berlin Heidelberg GmbH
A Springer-Verlag Company

Professor Lakhmi C. Jain
Knowledge-Based Intelligent
Engineering Systems Centre
University of South Australia
Adelaide, Mawson Lakes
South Australia 5095
lakhmi.jain@unisa.edu.au

Professor Beatrice Lazzerini
Dipartimento di Ingegneria della Informazione
University of Pisa
Via Diotisalvi, 2
56126 Pisa
Italy
beatrice@iet.unipi.it

Professor Ugur Halici
Department of Electrical Engineering
Middle East Technical University
06531 Ankara
Turkey
halici@metu.edu.tr

ISSN 1434-9922

DOI 10.1007/978-3-7908-1857-4

Cataloging-in-Publication Data applied for
Die Deutsche Bibliothek – CIP-Einheitsaufnahme
Innovations in ART neural networks: with 30 tables / Lakhmi C. Jain; Beatrice Lazzerini; Ugur Halici (eds.). – Heidelberg; New York: Physica-Verl., 2000
(Studies in fuzziness and soft computing; Vol. 43)

Originally published by Physica-Verlag Heidelberg in 2000.
MyCopy version of the original edition 2000

Cover design: Erich Kirchner, Heidelberg

88/2202-5 4 3 2 1 0 – Printed on acid-free paper
www.springer.com/mycopy

DEDICATION

This book is dedicated to Stephen Grossberg and Gail Carpenter.

PREFACE

At the dawn of the new millenium, the design and application of artificial neural networks continues to be the subject of intense research both in industry and in universities. In the last two decades, artificial neural networks have been refined and widely used by researchers and application engineers. The Adaptive Resonance Theory (ART) neural network has undergone evolution more than any other artificial neural network. The ART network remains plastic, or adaptive, in response to significant events and yet remains stable in response to irrelevant events. This stability-plasticity property is a great step towards realizing intelligent machines capable of autonomous learning in real time environment.

This book presents but a sample of the research on the evolution of ART neural network and its applications. Interested readers will find many more innovations in ART such as FUZZY ART, ART2, ART2-A, ART3, ARTMAP, ARTMAP-PI, ARTMAP-DS, Gaussian ARTMAP, EXACT ART, and ART-EMAP.

This book consists of eight chapters. The first chapter, by Filippidis, Russo and Jain, presents a novel extension of ART2 and its application in surface landmine detection. The ART2 is extended to provide a fuzzy output value, which indicates the degree of familiarity of a new analogue input pattern to previously stored patterns in the long term memory of the network. This extension is demonstrated to result in improved real-time landmine detection capabilities.

In the second chapter, Blume presents an alternative method for computing the FUZZY ART learning and classification functions. The vigilance criteria is checked before comparing the choice functions. This re-ordering of operations reduces the computation complexity. It is shown that this approach leads to a simple neural architecture that is better suited to parallel implementation.

The third chapter, by Blume and van Blerkom is on FUZZY ARTMAP modifications for intersecting class distributions. It is shown that the standard FUZZY ARTMAP algorithm over-trains and provides poor generalization on data drawn from intersecting class distributions. A modified FUZZY ARTMAP algorithm is presented to overcome these limitations. The performance of the improved algorithm is demonstrated using examples from speech and image understanding domains.

The fourth chapter, by Tan, demonstrates that the supervised ART architecture is compatible with IF-THEN rule-based symbolic representation. Specifically, the knowledge learned by a supervised ART system can be readily translated into rules for interpretation.

The fifth chapter, by Bartfai and White, introduces two ART-based networks that are capable of learning stable hierarchical clusterings of arbitrary input sequences by self-organization. The networks are cascades of ART modules, and implement clustering by successively "joining" (HART-J) or "splitting" (HART-S) the original input. Properties of both networks are shown, along with experimental results to demonstrate the representational capabilities of both networks and compare their performance to other hierarchical clustering algorithms. The Fuzzy HART-S network and optimization methods to HART-S are also discussed.

In the sixth chapter, Lim and Harrison describe the design of novel ART-based intelligent systems that are able to learn, and, at the same time, refine their knowledge. FUZZY ARTMAP and the probabilistic neural network are integrated to form a hybrid system that possesses the desirable properties for incremental, casual learning as well as for Bayesian probability estimation.

In the seventh chapter, Lim and Harrison, present an evaluation of the practical applicability of the classifier system described in the previous chapter. Two medical decision-making problems comprising real patient records are used to demonstrate the merits of the classifier system.

The last chapter by Serrano-Gotarredona and Linares-Barranco presents the implementation of ART1 systems. The enhanced sets of mathematical algorithms are used to optimize the design of hardware. It is demonstrated that this approach offers significant speed improvements over the conventional hardware implementation approaches.

This book will be useful to researchers, practicing engineers, and students who are interested in the novel extension of ART neural networks and their applications.

We are grateful to the authors for their contributions. We acknowledge the assistance of Berend Jan van der Zwaag, Ashlesha Jain, and Ajita Jain in the preparation of the manuscript. We thank Professor Janusz Kacprzyk for the opportunity to publish this book, and the Springer-Verlag Company for their excellent editorial assistance.

Lakhmi Jain, Australia
Beatrice Lazzerini, Italy
Ugur Halici, Turkey

CONTENTS

CHAPTER 1

NOVEL EXTENSION OF ART2 IN SURFACE LANDMINE DETECTION

A. Filippidis
Land Operations Division
Defence Science and Technology Organisation
P.O. Box 1500 Salisbury S.A. 5108, Australia

M. Russo
Dept. of Physics, University of Messina
Contrada Papardo - Salita Sperrone, 31
98166 Sant'Agata (ME), Italy

L.C. Jain
Knowledge Based Intelligent Engineering Systems Center
University of South Australia
Adelaide, Mawson Lakes, S.A., 5095, Australia

The Adaptive Resonance Theory 2 (ART2) neural network architecture is extended to provide a fuzzy output value, which indicates the degree of familiarity of a new analogue input pattern to previously stored patterns in the long term memory of the network. The outputs of the multilayer perceptron and this modified ART2 provide an analogue value to a fuzzy rule-based fusion technique which also uses a processed polarisation resolved image as its third input. In real-time situations these two classifier outputs indicate the likelihood of a surface landmine target when presented with a number of multispectral and textural bands. Due to the modifications in ART2, this updated alternative architecture has improved real-time landmine detection capabilities.

1 Introduction

There is a tremendous world-wide interest in the design and applications of artificial neural networks [1] both in universities and in industry. In the last two decades the artificial neural networks have been refined and widely used by the researchers and application engineers. We have not witnessed such a large degree of evolution in any other artificial neural network as in the Adaptive Resonance Theory (ART) neural network. We have extended the Adaptive Resonance Theory 2 (ART2) neural network architecture to allow it to provide a fuzzy input into a fuzzy rule based fusion system as shown in Figure 2. A familiarity flag which indicates the degree of match of a new analogue input pattern to previous patterns stored in the long term memory weights of the network has been added to indicate the likelihood of a surface landmine with a value ranging from zero (non-surface landmine) to one (surface landmine). The outputs of a multilayer perceptron together with the output of the modified ART2 provide two of the three input mine identity attributes (A and B) to the fusion system shown in Figure 4.

We have investigated the detection of surface land mines, given multiple registered images of the mined area obtained from a suite of visible to infrared wavelength sensors. The novel approach takes the outputs from two different imaging sensors; a thermal (infrared) imager fitted with a rotating polariser and a multispectral (optical) camera. The target information from the two images is fused together using a fuzzy rule based system, which uses a novel approach to fuzzify ART2's output. Compared to the earlier commercial system [2], the new approach is more suited to above surface landmines because (a) the multispectral sensor will only identify surface targets; and (b) the rotating polarisation filter attached to the thermal imager lens will now highlight surface targets by its particular polarisation signature. The technique reported here is less complicated in terms of the number of processing stages, uses the modifications in ART2 to improve its real-time capability and only well known preprocessing techniques are used.

2 Adaptive Resonance Theory

Adaptive resonance architectures are neural networks that self-organise stable recognition codes in real time in response to arbitrary sequences of input patterns. The basic principles of adaptive resonance theory (ART) were introduced in [3]. ART2 networks self-organise stable recognition categories in response to arbitrary sequences of analogue (gray-scale, continuous-valued) input patterns, as well as binary input patterns.

ART networks encode new input patterns, in part, by changing the weights, or long term memory (LTM) traces, of a bottom-up adaptive filter (refer to Figure 1). This filter is contained in pathways leading from a feature representation field (F1) to a category representation field (F2) whose nodes undergo cooperative and competitive interactions. Such a combination of adaptive filtering and competition, sometimes called competitive learning, is shared by many other models of adaptive pattern recognition and associative learning. In ART network, however, it is a second, top-down adaptive filter that leads to the crucial property of code self-stabilisation. Such top-down adaptive signals play the role of learned expectations in an ART system. They enable the network to carry out attentional priming, pattern matching, and self-adjusting parallel search.

As shown in Figure 1, the fields F1 and F2, as well as the bottom-up and top-down adaptive filters, are contained within ART's attentional system. An auxiliary orienting subsystem becomes active when a bottom-up input to F1 fails to match the learned top-down expectation read-out by the active category representation at F2. In this case, the orienting subsystem is activated and causes rapid reset of the active category representation at F2. This reset event automatically induces the attentional subsystem to proceed with a parallel search. Alternative categories are tested until either an adequate match is found or a new category established. The search remains efficient because the search strategy is updated adaptively throughout the learning process. The search proceeds rapidly, relative to the learning rate. Thus significant changes in the bottom-up and top-down adaptive filter occurs only when a search ends and a matched F1 pattern resonates within the system.

The fact that learning within an ART system occurs only within resonant state enables such a system to solve the design trade off between plasticity and stability. Plasticity, or the potential for rapid change in the LTM traces, remains intact indefinitely, thereby enabling an ART architecture to learn about future unexpected events until it exhausts its full memory capacity.

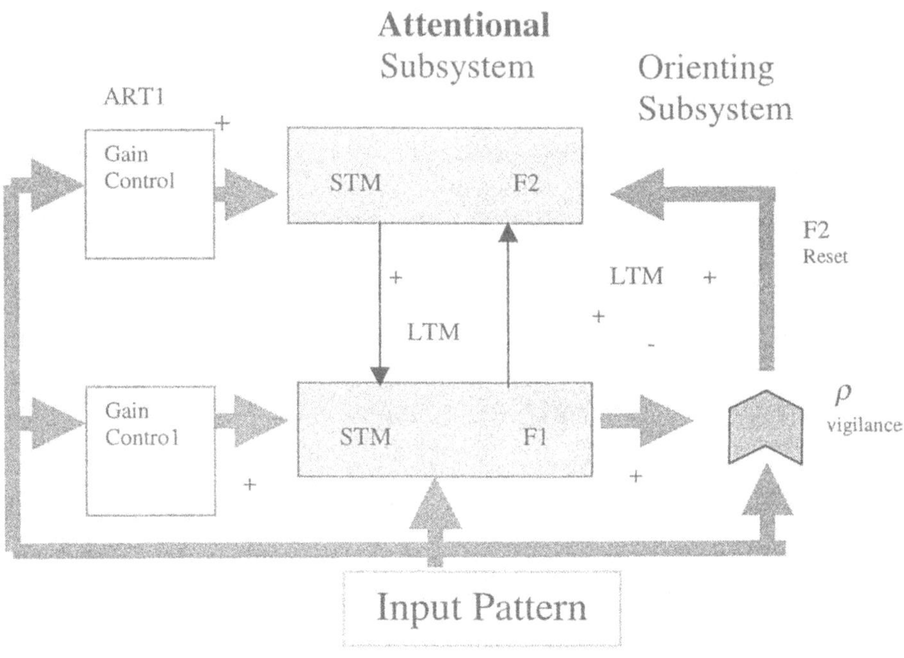

Figure 1. A typical ART1 architecture. Rectangles represent fields where STM patterns are stored. Thin arrows represent adaptive filter pathways and thick arrows represent paths which are not adaptive. Squares represent gain control nuclei, which sum input signals. Gain control at F1 and F2 coordinates STM processing with input presentation.

The architectures adaptive search enables it to discover and learn appropriate recognition codes without getting trapped into spurious states or local minima. The criteria for an adequate match between an input pattern and a chosen category template is adjustable in an ART architecture. The matching criterion determined by the vigilance parameter that controls activation of the orienting subsystem. All other things being equal, higher vigilance imposes a stricter matching criterion, which in turn partitions the input set into finer categories. Lower vigilance tolerates top-down/bottom-up mismatches at F1, leading inturn to course categories.

The top-down learned expectations and the orienting subsystem are not the only means by which an ART network carries out active regulation of the learning process. Attentional gain control at F1 and F2 also contribute to this active regulation. Gain control nuclei are represented as large filled circles as shown in Figure 1. This network paradigm is widely referred to as ART1.

3 Evolution of Adaptive Resonance Theory

Such a tremendous evolution is not witnessed in any other artificial neural network as in the Adaptive Resonance Theory (ART) neural network. Although self-organised recognition of binary patterns is useful in many applications including the recognition of written text, many other applications require to categorize analogue input patterns. Grossberg and Carpenter have developed a class of architectures, generally called ART2, for this purpose.

On the surface, ART2 differs from ART1 (shown in Figure 1) only in the nature of the input patterns: ART2 accepts analog (or grey scale) vector components as well as binary. This capability represents a significant enhancement to the system. Beyond the surface difference between ART1 and ART2 lie architectural differences that give ART2 its ability to deal with analog patterns. These differences are sometimes more complex, and sometimes less complex, than the corresponding ART1 structures. Aside from the obvious fact that binary and analogue patterns differ in nature of their respective components, ART2 must deal with additional complications. For example, ART2 must be able to recognise the underlying similarity of identical patterns superimposed on constant backgrounds having different levels. Compared in absolute sense, two such patterns may appear entirely different when, in fact, they should be classified as the same pattern.

Following is a list of ART paradigms widely reported in the literature.

- ART1: Self-organised recognition of binary input patterns
- FUZZY ART: Incorporates computations from fuzzy set theory into the ART1.
- ART2: Self-organised recognition of analogue (including binary) input patterns
- ART2-A: Efficient ART2
- ART3: For parallel search, or hypothesis testing of distributed recognition codes in a multilevel network hierarchy.
- ARTMAP: Autonomously learns to classify arbitrarily many, arbitrarily ordered vectors into recognition categories based on predictive success.
- ARTMAP-PI: Extension of ARTMAP to provide probabilistic estimates of outcomes.
- ARTMAP-DS: Extension of ARTMAP to discriminate between similar inputs by discounting similarities.
- Gaussian ARTMAP: Synthesis of Gaussian classifier and ART for incremental supervised learning of analogue multidimensional maps.
- EXACT ART: Complete implementation of ART including all regulatory and logical functions.
- ART-EMAP: A new class of neural network architecture for the recognition of pattern classes after supervised and unsupervised learning.
- MART: An ART-based neural network for adaptive classification of multichannel signal patterns without prior supervised learning.

4 ART2 Familiarity Flag

The architecture, familiarity ART2 shown in Figure 2 is used specifically in a supervised fashion. Unlike other ART networks once trained and weights stored in the long term memory of the network it has the added unique ability to flag itself using a single output (using the familiarity flag shown in Figure 2) that it has come across the new input pattern before, and hence this single output can conveniently be directly fed into the fusion architecture shown in Figure 4. Hence at a certain vigilance value the familiarity flag will also indicate the degree of familiarity of a new input vector to the closest previous pattern stored in the long term memory of the network. The novel modification of the ART2 architecture enhances the automatic recognition system by allowing it to now process the information and output the likelihood of a surface landmine target automatically in a format suitable to the fusion engine, when presented with multispectral and textural information. The main advantages of ART2 over the MLP are its on-line training which is a desirable robust feature the automatic target recognition system will need if the external environment variables change drastically from the initial training set.

In fuzzy ART [4] the fuzzy logic AND connective, min, is used to extend the method to real values in ART1. This formulation of similarity is unusual and according to reference [5] inconsistent, since the resulting measure is not the canonical Norm 1 metric, but is used here in order to maintain the biological inspiration of ART. Field *et al.* in reference [5] also mentioned that a problems encountered in his experimental results of using the Fuzzy ART network was apparent when the weight vectors of the classifying units were found to be very small. A comparison of Fuzzy ART to other unsupervised techniques in reference [5] also identified several clustering inconsistencies in their examples due to the Fuzzy Art similarity measure used. Due to the above negative results and concerns by the authors in the journal described in reference [5], and due to the small amount of effort required to modify ART2, whose original (unmodified) architecture we had used successfully with good results in experiments described in journal reference [6], the modified ART2 architecture was used here, as shown in Figure 2.

ART2's basic architecture taken from [7] is shown in Figure 2, together with the addition of a familiarity flag to the right hand side of the dotted line. The familiarity flag indicates the degree of familiarity of a new input vector *Ii* to previous patterns stored in long term memory zij, zji. The familiarity flag now automates ART2 so it can automatically indicate the degree of familiarity and can be directly fed into the ATR fuzzy fusion system hence automating the whole system. The open arrows in Figure 2 indicate specific patterned inputs to target nodes and the filled arrows indicate non specific gain control units. The F1 layer is divided into 6 sub-layers *wi, xi, ui, vi, pi,* and *qi*. The gain control nuclei shown by (large filled circles) non specifically inhibit target nodes in proportion to the L2-norm of short term memory (STM) activity in their source fields (equations (3), (4), (7), (8) and (9)). When F2 makes a choice, $g(yj)=d$ if the jth F2 node is active and $g(yj)=0$ otherwise. The orienting system resets F2 layer for a given vigilance ρ.

A brief summary [7] of the asymptotic solutions and initial parameter constraints to the dynamic equations, and the fast learning mode used to implement the standard ART2 network shown on the left hand side of Figure 2 is described below.

$$a,b > 0, \qquad 0 \le d \le 1, \qquad \frac{cd}{1-d} \le 1,$$
$$0 \le \theta \le 1, \qquad 0 \le \rho \le 1, \qquad e << 1.$$

Modified top-down and bottom-up weights:

$$zij, zji = \frac{ui}{1-d} \tag{1}$$

The dimensionless equations below, that characterise the STM activities, *pi, qi, vi, ui, wi,* and *xi,* computed at F1:

$$pi = ui + \sum_j g(yj)zji \tag{2}$$

$$qi = \frac{pi}{e + \|p\|} \tag{3}$$

$$ui = \frac{vi}{e + \|v\|} \tag{4}$$

$$vi = f(xi) + bf(qi) \tag{5}$$

$$wi = Ii + aui \tag{6}$$

$$xi = \frac{wi}{e + \|w\|} \tag{7}$$

where $\|V\|$ denotes the L2-norm of a vector *V*, *yj* is the STM activity of the *jth*. F2 node, and *zji* is the long term memory (LTM) weight in the pathway from the *jth*. F2 node to the ith. F1 node. Where *i* and *j* are the number of units in each F1 and F2 layers respectively. The nonlinear signal function *f* in equation (5) is typically of the form:

$$\begin{aligned} f(x) &= 0 \quad && if \quad 0 \le x < \theta \\ f(x) &= x \quad && if \quad x \ge \theta \end{aligned}$$

The degree of match between a STM pattern at F1 and an active LTM pattern is determined by the vector *r=(r1 ... ri).*

$$ri = \frac{ui + cpi}{e + \|u\| + \|cp\|} \tag{8}$$

The orienting subsystem is assumed to reset F2 whenever an input pattern is active and

$$\frac{\rho}{e + \|r\|} > 1 \tag{9}$$

The additional familiarity flag will indicate 1 if there is prior memory recall and a value of zero up to a value of one as the degree of the input vector matches the stored vectors in LTM. Modified equations used to fuzzify ART2's outputs for this particular application are as follows:

$$qi' = \frac{|pi - qi|}{\|pi - qi\|} \tag{10}$$

Taking the cosine of the angle between *qi'* and the normalised input vector *Ii* as shown on the right hand side of Figure 2 gives the equation for the familiarity flag:

$$familiarity\_flag = \cos(qi', Ii) = \frac{\rangle qi', Ii \langle}{e + \|qi'\| \ \|Ii\|} \tag{11}$$

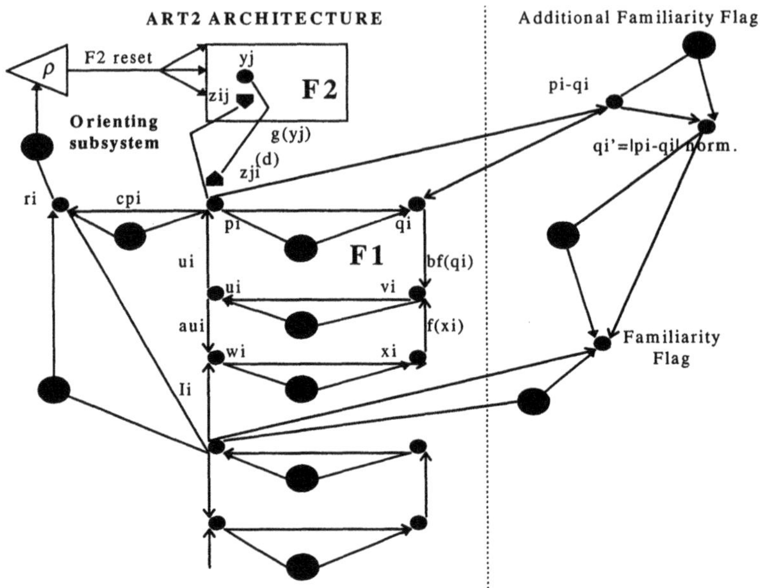

Figure 2. ART2 architecture, together with the addition of a familiarity flag (which indicated degree of familiarity of a new input vector Ii to previous patterns stored in long term memory zij, zji) to the right of the dotted line. Open arrows indicate specific patterned inputs to target nodes. Filled arrows indicate non specific gain control units. The F1 layer is divided into 6 sublayers wi, xi, ui, vi, pi, and qi. The gain control nuclei shown by (large filled circles). Orienting system resets F2 layer for a certain vigilance ρ.

5 Test Mine Field

For this study a small test mine field was established. The field consisted mainly of seasonal grasses with a number of walking tracks through it. The soil was hard packed red loam. Figure 3 is a colour composite image of the test field, and shows three mines (A, B, C in circles) VS50, M14 and PMM with a surrogate landmine (D, made of a polystyrene block similar to the material of the M14 mine) together with man-made clutter (E, F, G, H, I in squares) such as a marker1, aluminium can, teflon block, marker2 and aluminium block.

The land mines were an American M14, an Italian VS50 anti-personnel mine, and a Russian PMM anti-tank mine Pucks of similar dimensions to the M14 made of teflon, PVC, polystyrene (PS) and aluminium were also used [8]. The explosive material was removed from the landmines in this experiment. The M14, VS50 and PMM mines have the following characteristics:

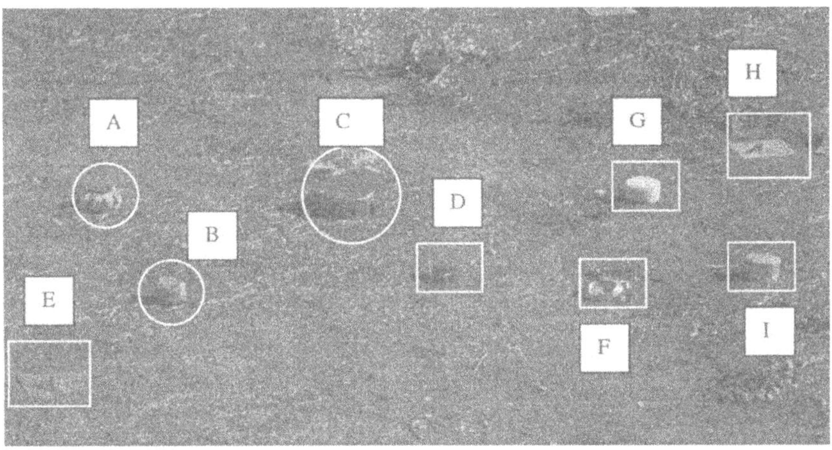

Figure 3. Composite colour image shows three mines (A, B, C in circles) VS50, M14 and PMM respectively, and one surrogate mine (D, made of a polystyrene block similar to the material of the M14 mine) together with man-made clutter (E, F, G, H, I in squares) such as a marker1, aluminium can, teflon block, marker2 and aluminium block respectively surrounded by dry grasses.

The Agema Thermovision 900 radiometric dual band thermal imaging system formed the basis of the data collection and recording of the polarisation images. Both the 3-5 micron and 8-12 micron scanners were available for use in this system, although only the 8-12 micron polarised images were used in this experiment. The rotating polariser consists of an Optometrics wire grid polarising filter which has precisely spaced grooves directly into a highly polished ZnSe substrate together with a small 12V electric motor which rotates the lens when mounted on the Agema scanners (8-12 micron). The Agema system was set up in the van adjacent to the test mine field. The heads of the thermal imaging scanners together with the digital camera were set up on a tripod and mounted on the roof of the van. The scanners were fitted with a rotating polarisation filter, which at the time of the experiment rotated at approximately 1 revolution a second. Software on a personal computer could stop the rotation at 0, 45, 90 and 135 degrees. The scanning height was 3.5 metres, and the surface mine field was located approximately 8 metres away. The digital camera recorded a number of images in the near IR spectrum (the bands were 449, 450, 550, and 600nm) and a colour composite image of the mine field. The Agema recorded the same minefield through a complete revolution of the rotating polariser (i.e., 0-360 degrees).

6 Automatic Detection of Surface Landmine

The automatic target recognition (ATR) system (shown in Figure 4) uses fuzzy rule based fusion to combine complementary information derived from both sensors to produce an output image showing the likelihood of mine locations. The two classifier inputs to the fusion process are the real-time output from ART2 and the MLP (shown as attribute B and A in Figure 4 respectively). The third input identity attribute "area" is the number of pixels in a 8x8 window (moving across the image one pixel at a time) calculated from the output of the processed IR polarisation image as shown in Figure 4, using a background discrimination algorithm [9].

A Neural Network classifier has been trained to discriminate against clutter and automatically detect land mines. One of the main reasons for concentrating on the neural network classifier, the MLP in particular, is because it has been shown [10] to perform equally as well as the classical Bayesian classifiers for classes of multispectral data that are normally distributed. Moreover, where non-normally distributed multispectral data is concerned, as is the case when classes of different areas or targets are combined as one class (such as three different mines), the MLP performs better with an increase in accuracy of at least 5% [10].

ART2 was selected for its robust ability to classify targets in real-time automating the system unlike the MLP which requires hours of training. The additional input of ART2 into the fuzzy fusion process has increased the detection rate as one of the surrogate landmines (D in Figure 3), was not detected by the MLP alone. ART2's inputs (using a 5×5 pixel window moving across each of the 4 bands shown in Figure 4) are the average spectral characteristics from the red, green and blue bands together with the three texture measures [11] contrast, second angular momentum, and correlation) derived from the 449nm band.

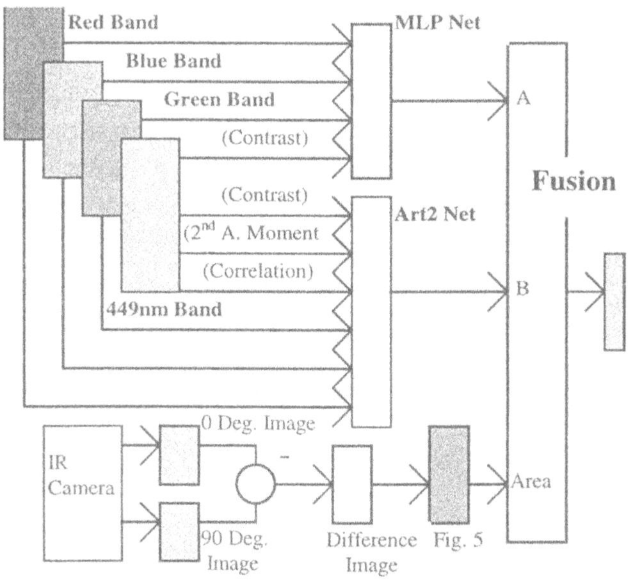

Figure 4. The ATR system uses a fuzzy rule based fusion technique to combine landmine identity attributes from the outputs of the MLP, ART2 and pre-processed thermal polarisation image.

Using the accuracy of test and training of the mine and non mine data a genetic algorithm (GA) tool [12] is used to find the optimum structure and inputs of the MLP neural network. The MLP uses the three spectral bands red, green and blue together with the textural measure contrast [8] derived from the 449nm band. Two polarisation resolved images are subtracted at two different polarisation angles (0 and 90 degrees). It is then processed using a background discrimination algorithm [9] to identify the man made targets in the image shown in Figure 5.

Polarisation [13] sensing technique used exploits the polarimetric thermal IR features of the unscattered and emitted electromagnetic radiation from the mines and the surrounding clutter. The fact that mines are man-made and are of deterministic geometric regularities, versus the surrounding randomly oriented media results in a high probability of detection and low false alarms when using thermal IR polarisation features. The polarisation resolved image on its own cannot identify the landmine targets, as it will detect most surface targets, but not distinguish them. The classifiers have the potential to identify targets but produce a very large number of misclassifications.

The digital camera provided several bands in the near infrared (449, 450, 550, and 600nm) and a colour composite image as shown in Figure 3. Different combinations of the following bands using three texture parameters computed from the Co-Occurrence Matrices [11] (second-angular momentum, correlation and contrast of an 8x8 window), together with their spectral values were explored using the MLP and ART2 Neural Networks. The implementation of the fuzzy rule based system involves a fusion equation [14] to combine identity attributes from ART2, MLP and pre-processed thermal polarisation image.

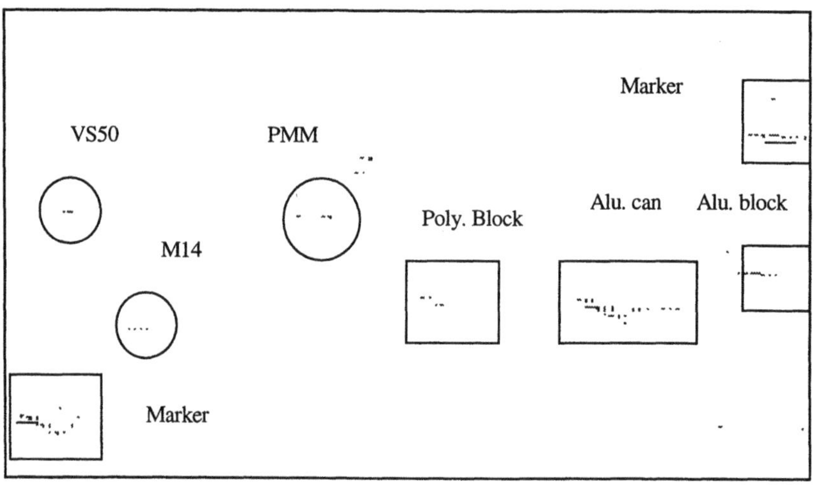

Figure 5. Processing the thermal polarisation images we were able to successfully identify 8 of the 9 man made targets three of which were mines (shown in circles) with only 4 false alarms.

The ATR system in Figure 4 is designed to use either one or both neural classifiers together with the polarisation resolved image obtained from the IR camera. The imaging system on its own cannot discriminate between landmines and other manmade targets. When fusing either one or both of the neural classifier outputs with the polarisation resolved image a dramatic reduction in false alarm rate (FAR) is obtained.

7 Conclusion

In this modified ATR system we have incorporated novel changes in the ART2 architecture enabling it to produce the fuzzified output needed in the fuzzy fusion stage. Sensor fusion has overcome the limitations of the individual sensors and processing techniques for the surface mine detection task. When one sensor or processing technique failed to provide all the necessary information as was very much the case in our system, a complementary sensor or processing technique (i.e classifier output) provided additional information which reduced our overall false alarm rate.

References

[1] Jain, L.C. and Vemuri, R. (Editors), (1998), *Industrial Applications of Neural Networks*, CRC Press USA.

[2] Bower, M., Cloud, E., Duvoisin, H., Long, D., and Hackett, J. (1995), "Development of automatic target recognition for infrared sensor-based close range land mine detector," *Martin Marietta Technologies Inc. Tech. Report*, P.O. BOX 555837, MP-040, Orlando FL 32812-5837.

[3] Carpenter, G.A. and Grossberg, S. (1987), *Pattern Recognition by Self-organising Neural Networks*, Academic Press Inc., pp. 399-410.

[4] Carpenter, G.A., Grossberg, S., and Rosen, D.B. (1991), "Fuzzy art: fast stable learning and categorization of analogue patterns by adaptive resonance systems," *Neural Networks*, vol. 4, no. 6, pp. 759-771.

[5] Field, S., Davey, N., and Frank, R. (1996), "Using neural networks to analyse software complexity," *Australian Journal of Intelligent Information Processing Systems*, vol. 3, no. 3, pp. 14-31, Spring.

[6] Filippidis, A. and Jain, L.C. (1997), "Identity attribute information in a multi-band aerial photo image using neural networks to automatically detect targets," *International Journal of Knowledge - based Intelligent Engineering Systems*, vol. 1, no. 1., January.

[7] Carpenter, G. and Grossberg, S. (1987), *Pattern Recognition by Self-Organising Neural Networks*, Academic Press Inc., pp. 399-410.

[8] Fuelop, K. and Hall, J. (1996), *Thermal Infrared landmine detection*, Technical Report DSTO-TR-0295 AR-009-485, pp. 3-19, January.

[9] Stacy, N., Smith, R., and Nash, G. (1994), *Automatic target recognition for the INGARRA airborne radar surveillance system*, D.S.T.O. Microwave Radar Division Internal Report, pp. 1-12, August.

[10] Whitbread, P.J. (1992), *Multispectral texture*, Ph.D. Thesis, The University of Adelaide, October.

[11] Harlick, R.M., Shunmuhham, K., and Distein, I. (1973), "Textural features for image classification," *IEEE Transactions, Man and Cybernetics*, vol. SMC-3, no. 6, November.

[12] "Neuro-genetic optimiser version 32202," BioComp Systems Inc. 2871, 152nd. Avenue N.E. Redmond, WA 98052.

[13] Ben-Dor, B., Oppenheim, U.P., and Balfour, L.S. (1992), "Polarisation properties of targets in backgrounds in the infrared," *SPIE*, vol. 1971, 8th. Meeting on Optical Engineering in Israel, pp. 68-77.

[14] Abdulghafour, M.B. (1992), *Data fusion through fuzzy reasoning applied to feature extraction from multi-sensory images*, Ph.D. Thesis, The University of Tennessee, Knoxville, pp. 41-96, December.

CHAPTER 2

AN EFFICIENT MAPPING OF FUZZY ART ONTO A NEURAL ARCHITECTURE

M. Blume
HNC Software, Inc., San Diego
U.S.A.

It is possible to eliminate the reset/resonance cycle of many of the ART algorithms without modifying their learning or classification rules. The enabling principle is to check the vigilance criterion before comparing the choice functions, rather than vice versa. This re-ordering of operations reduces the computational complexity. It leads to a simpler neural architecture that is better suited to parallel implementation and facilitates understanding the algorithm.

1 Introduction

The adaptive resonance theory [1], [2] models feedback and recurrent connections in natural neural systems. The dynamics of the ART algorithms (e.g. ART1 [2], ART2 [3], and Fuzzy ART [4]) approximate those described by the equations that constitute the theory. Like the systems that inspired them, the algorithms feature self-organization and learned pattern recognition. However, the recurrent connections and resonance that are central to the theory are awkward and unnecessary [5], [6], [7], [8] in many of the ART algorithms.

This chapter describes an alternative method for computing the Fuzzy ART learning and classification functions. The resulting weight values and class assignments are identical to those of the original algorithm. However, because the order of the operations is altered, the resonance/ reset cycle is eliminated. The time required for the computations is fixed, rather than depending on the relationships between the input vector and the weight templates. This reduces the worst-case complexity of the computation.

When mapped onto a neural architecture, the standard approach requires bi-directional synapses, weight transport, or weight duplication – the architecture described here does not. Furthermore, the proposed architecture consists of one fewer layer of processing elements than the original Fuzzy ART architecture [9]. These characteristics facilitate parallel hardware implementation of the method.

This chapter focuses on Fuzzy ART. However, the modified data flow is equally applicable to other ART algorithms, including ART1, ART2, and Fuzzy ARTMAP. The mapping onto a simpler neural architecture serves as a tool for envisioning and understanding any of the ART algorithms.

2 Traditional Implementation

Fuzzy ART clusters vectors based on two separate distance criteria, *match* and *choice*. For input vector $\boldsymbol{I}$ and category j, the match function is defined by

$$S_j(\boldsymbol{I}) \equiv \frac{|\boldsymbol{I} \wedge \boldsymbol{w}_j|}{|\boldsymbol{I}|} ,$$

where $\boldsymbol{w}_j$ is an analog-valued weight vector associated with cluster j. $\wedge$ denotes the fuzzy AND operator, $(\boldsymbol{p} \wedge \boldsymbol{q})_i \equiv min(p_i, q_i)$, and the norm $|\cdots|$ is defined by $|\boldsymbol{p}| \equiv \sum_i |p_i|$. The choice function is defined by

$$T_j(\boldsymbol{I}) \equiv \frac{|\boldsymbol{I} \wedge \boldsymbol{w}_j|}{\alpha + |\boldsymbol{w}_j|} ,$$

where α is a small constant. Each input vector is assigned to the category that maximizes $T_j(\boldsymbol{I})$ while satisfying $S_j(\boldsymbol{I}) \geq \rho$, where the *vigilance*, ρ, is a constant, $0 \leq \rho \leq 1$.

Carpenter, et al. [4] propose searching for the category J which maximizes $T_j(\boldsymbol{I})$ and then checking whether $S_J(\boldsymbol{I}) \geq \rho$. If not, category J is marked as ineligible (*reset*) and the search is repeated until a satisfactory category is found (*resonance*). This procedure is illustrated in Figure 1.

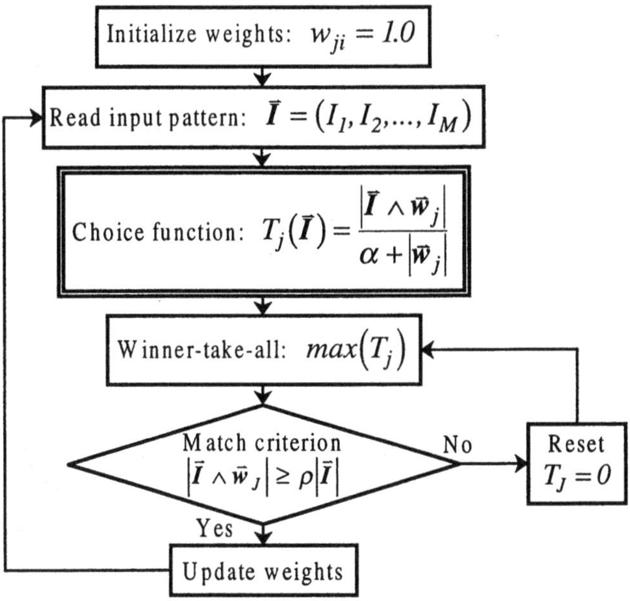

Figure 1. Flowgraph representation of the original Fuzzy ART algorithm. The choice function is computed for all categories, and the match criterion is checked only for the category with the largest choice value. If this category does not satisfy the match criterion, it is marked ineligible (reset), and the process is repeated.

For M inputs and N clusters, the total number of weights is MN. Consequently, the computational complexity of computing $|\boldsymbol{I} \wedge \boldsymbol{w}_j|$ for all clusters j is $O(MN)$. Finding $argmax(T_j)$ is $O(N)$. Upon reset, $argmax(T_j)$ must be re-computed with the previous winners disabled. Since up to N resets (one per category) may be required for a single input vector, the worst-case complexity of finding $argmax(T_j \mid S_j(\boldsymbol{I}) \geq \rho)$ via this scheme is $O(N^2)$, and the worst-case complexity of the entire procedure is $O(MN) + O(N^2)$.

The reset/resonance cycle is realized in the three-layer neural network illustrated in Figure 2. Calculation of the choice function takes place in layer F_2, whereas calculation of the match function and comparison to the vigilance parameter takes place in layer F_1 and at the reset node. Since the weight values, w_{ji}, are required for both calculations, copies are stored and independently updated in each of the two layers. The length of time between input presentation and selection of the corresponding cluster is variable, depending on how many search cycles are required.

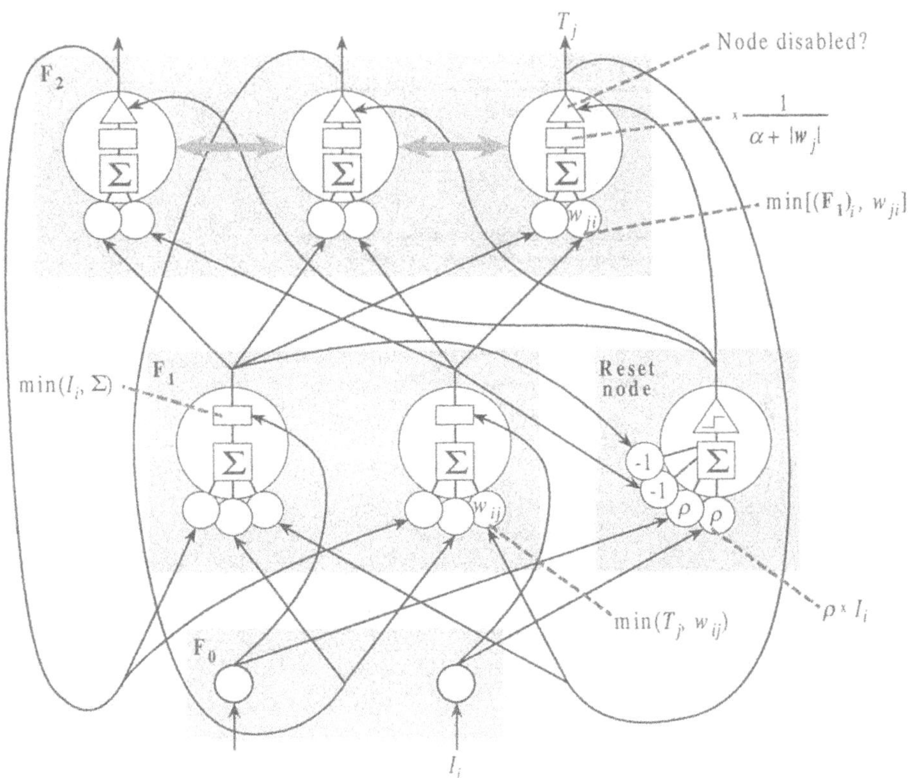

Figure 2. The Fuzzy ART architecture proposed by Carpenter, Grossberg, and Rosen [9] consists of three layers of processing elements and utilizes two identical sets of weights. Some control signals are not shown.

3 Testing the Vigilance Before Comparing Choice Values

An alternative approach, illustrated in Figure 3, is to check the match criterion for every cluster and then compare the choice functions of only those clusters satisfying the match criterion. As in the original scheme, the winning category is specified by $argmax(T_j \mid S_j(\boldsymbol{I}) \geq \rho)$ - neither the weight values, nor the classification functions, nor the classification itself differ from the standard Fuzzy ART algorithm.

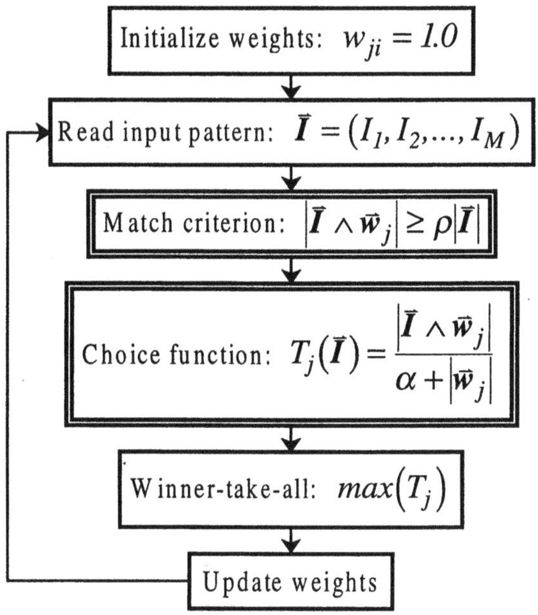

Figure 3. Flowgraph representation of the modified fuzzy ART algorithm. The match criterion is checked for all categories, and the choice function is computed only for those categories satisfying the match criterion. This sequence of the operations is different from that of the original algorithm (Figure 1), but the classification of input vectors remains unchanged.

Since the reset/resonance cycle is eliminated, $argmax(T_j)$ is only computed once per input vector at a cost in complexity of $O(N)$. Consequently, the computation of $|\boldsymbol{I} \wedge w_j|$ for all clusters j dominates, and the complexity of the entire procedure is $O(MN)$.

Figure 4 represents an architecture based on this order of the operations. The choice and match functions are both computed in the same layer. In a parallel implementation, both the match and choice functions can be computed simultaneously for all clusters, and execution of the algorithm takes a constant length of time per input vector. Finally, each processing element (PE) executes the same operations at every time step, eliminating the control signals associated with the reset/resonance process and facilitating pipelining.

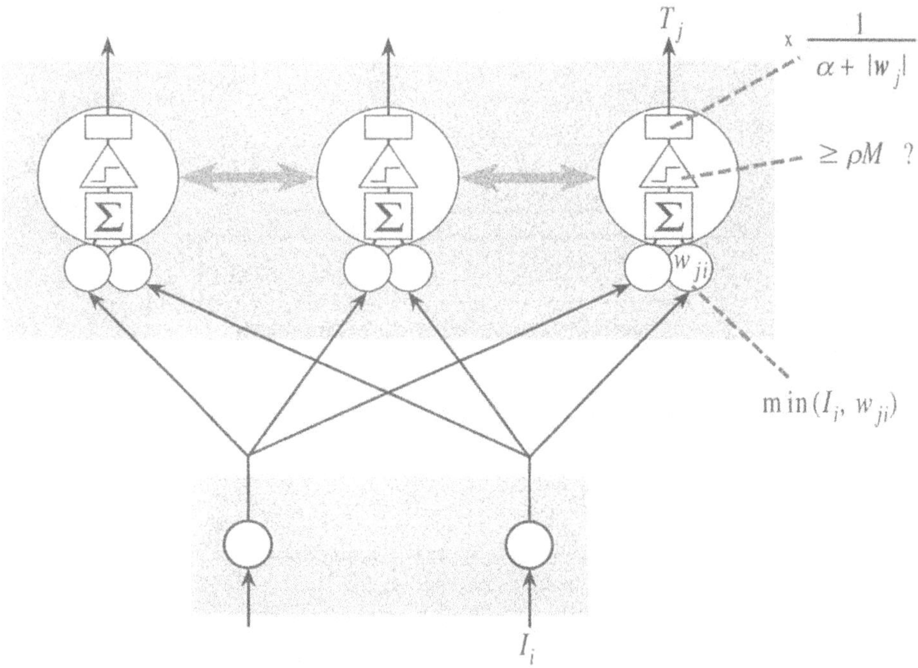

Figure 4. Simpler Fuzzy ART architecture, consisting of two layers and requiring only one set of weights. Complement coding of the input vectors, eliminating the need for computation of $|\boldsymbol{I}|$, is assumed. The winner-take-all interconnections (shaded arrows) could be replaced by one additional processing element that determines which layer 2 PE has the maximal output.

In the proposed architecture, every processing element must compute the match criterion. Fortunately, this computation requires minimal additional effort. The relationship $S_j(\boldsymbol{I}) \geq \rho$ is equivalent to $|\boldsymbol{I} \wedge \boldsymbol{w}_j| \geq \rho|\boldsymbol{I}|$. The quantity on the left hand side of this equation is needed for the computation of $T_j(\boldsymbol{I})$ in any case, and hence requires no additional calculations.

Typically, a complement coding scheme is utilized which guarantees that the L_1 lengths of all input vectors are identical [4], [10]. In this case, $\rho|\boldsymbol{I}|$ is a constant (the number of inputs, M), and the evaluation of the match criterion is reduced to a single comparison. If complement coding is not used, the magnitude of every input vector, $|\boldsymbol{I}|$, must be calculated. This may be accomplished by a single additional processing element (equivalent to the reset node in the original architecture) that transmits its output value to the PEs of the second layer.

Thus, the only additional effort over the original algorithm is the evaluation of the match criterion for all categories (rather than just the one with the highest choice value), which requires N comparisons. For the purpose of comparison, note that determining which $T_j(\boldsymbol{I})$ is maximal requires $N - 1$ comparisons with either method. In the standard formulation, this operation is carried out once per reset. Consequently, if the original scheme requires more than two search cycles on average, calculating the match criterion before the choice function reduces the computation time in a serial implementation.

The fuzzy ART learning rule is given by

$$\boldsymbol{w}_J^{new} = \beta(\boldsymbol{I} \wedge \boldsymbol{w}_J^{old}) + (1-\beta)\boldsymbol{w}_J^{old} ,$$

where $0 < \beta \leq 1$. This is equivalent to

$$w_{Ji}^{new} = \begin{cases} w_{Ji}^{old} & w_{Ji} \leq I_i \\ w_{Ji}^{old} - \beta\left(w_{Ji}^{old} - I_i\right) & w_{Ji} > I_i \end{cases} ,$$

In either architecture, the weight template of the winning category is updated using locally available information. However, in the proposed architecture, there is only one copy of the weights, further facilitating parallel implementation.

4 Summary

Two characteristics make Fuzzy ART an interesting target for implementation in parallel hardware. First, the algorithm specifies a comparison operation at the synapse level, rather than the more common (and, in digital electronics, more complicated) multiplication. Second, the algorithm converges well with as few as four bits of precision [11]. This chapter describes an architecture that eliminates a number of control signals and the weight replication that were previously suggested, resulting in a system which is especially well suited to massively parallel implementation.

The mapping onto a less complicated architecture promotes an intuitive understanding of the ART clustering process, making it easy to envision how each input vector is assigned to the cluster that maximizes the choice function while satisfying the match criterion.

The only change from the initial formulation of the algorithm is the order in which the vigilance test and the search for the maximum choice value are carried out. This reduces the worst-case complexity from $O(MN) + O(N^2)$ to $O(MN)$. The classification of the input vectors remains identical to that produced by the original procedure.

Acknowledgments

The author developed the algorithm and architectures described here while a graduate student in the Department of Electrical and Computer Engineering at the University of California, San Diego. This research was supported by the Advanced Research Projects Agency (ARPA F49620-92-J-0520).

References

[1] Grossberg, S. (1976), "Adaptive pattern classification and universal recoding: II. feedback, expectation, olfaction, and illusions," *Biologic Cybernetics*, vol. 23, pp. 187-202.

[2] Carpenter, G.A. and Grossberg, S. (1987), "A massively parallel architecture for a self-organizing neural pattern recognition machine," *Computer Vision, Graphics, and Image Processing*, vol. 37, pp. 54-115.

[3] Carpenter, G.A. and Grossberg, S. (1987), "ART2: Self-organization of stable category recognition codes for analog input patterns," *Applied Optics*, vol. 26, pp. 4919-4930.

[4] Carpenter, G.A., Grossberg, S., and Rosen, D.B. (1991), "Fuzzy ART: fast stable learning and categorization of analog patterns by an adaptive resonance system," *Neural Networks*, vol. 4, pp. 759-771.

[5] Blume, M. and Esener, S.C. (1995), "Optoelectronic fuzzy ARTMAP processor," *1995 OSA Technical Digest Series*, vol. 10, pp. 213-215.

[6] Serrano-Gotarredona, T., Linares-Barranco, B., and Huertas, J.L. (1995), "A real time clustering CMOS neural engine," *Advances in Neural Information Processing Systems*, vol. 7, pp. 755-760.

[7] Blume, M. and Esener, S.C. (1997), "An efficient mapping of fuzzy ART onto a neural architecture," *Neural Networks*, vol. 10, pp. 409-411.

[8] Burwick, T. and Joublin, F. (1998), "Optimal algorithmic complexity of fuzzy ART," *Neural Processing Letters*, vol. 7, pp. 37-41.

[9] Carpenter, G.A., Grossberg, S., and Rosen, D.B. (1991), *A Neural Network Realization of Fuzzy ART* (Technical Report CAS/CNS-91-021), Boston University, Center for Adaptive Systems, Boston, MA.

[10] Huang, J., Georgiopoulos, M., and Heileman, G.L. (1995), "Fuzzy ART properties," *Neural Networks*, vol. 8, pp. 203-213.

[11] Rubin, M.A. (1995), "Issues in automatic target recognition from radar range profiles using fuzzy ARTMAP," *Proceedings of the 1995 World Congress on Neural Networks*, pp. 197-202.

CHAPTER 3

FUZZY ARTMAP MODIFICATIONS FOR INTERSECTING CLASS DISTRIBUTIONS

M. Blume
HNC Software, Inc., San Diego
USA

D.A. Van Blerkom
Photobit Corporation, Pasadena
USA

As originally defined, the Fuzzy ARTMAP algorithm performs poorly with intersecting class distributions, as commonly occur in real-world data. This chapter describes several modifications which eliminate the underlying category proliferation problem. The performance of the original and modified algorithms is demonstrated with examples from the speech and image understanding domains.

1 Introduction

Classification is the process of assigning a data vector to one of several pre-defined groups. If the feature space cannot be subdivided into regions with individual class associations, the underlying class distributions are said to intersect. This is common in real problems due to noise, classification uncertainty and errors, the use of sub-optimal features, loss of context, and so on. Figure 1 shows an example.

Classification of points from intersecting class distributions has long been identified as a more difficult problem than classification of points from separable distributions [1]. It is not possible to classify a point lying in the region of overlap with 100% certainty. The best a system can do is estimate the probability that a training point belongs to a particular class, or equivalently determine a fuzzy value of the class membership of the data point [2]. If the output of a classifier is restricted to a single label, the best generic approach is to select the class from which the data point was most probably drawn.

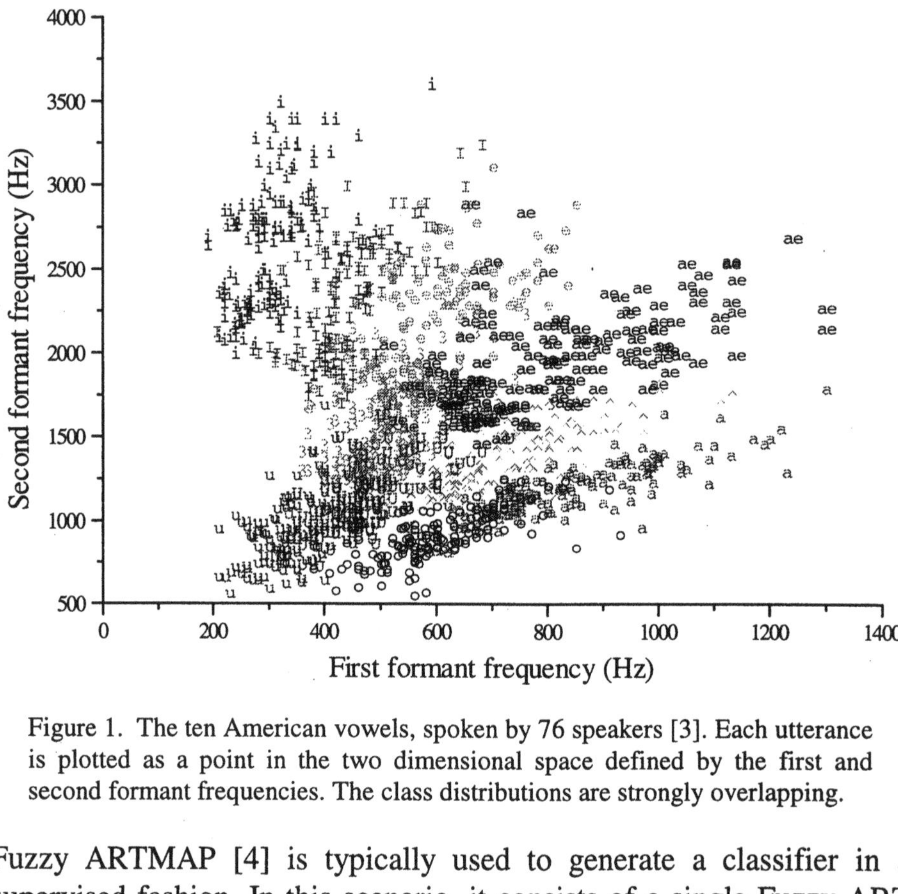

Figure 1. The ten American vowels, spoken by 76 speakers [3]. Each utterance is plotted as a point in the two dimensional space defined by the first and second formant frequencies. The class distributions are strongly overlapping.

Fuzzy ARTMAP [4] is typically used to generate a classifier in a supervised fashion. In this scenario, it consists of a single Fuzzy ART module and a MAP field. During learning, the system is presented with training vectors that have been labeled according to the class to which each belongs. For example, vectors of features extracted from short segments of speech might be labeled according to the phoneme being spoken during each segment. The Fuzzy ART module assigns the current training vector to a category. The MAP field associates a class with each category. If the class label of the training vector differs from the class associated with the chosen category, the MAP field causes the

Fuzzy ART search to be repeated, this time with attention restricted to smaller categories. If this procedure, termed *match tracking*, successfully assigns the training vector to a category with the correct class association, the weights defining that category are updated, making it more likely to capture that training vector in the future. If no category with the correct class association is found, a new category is created.

When clouds of data points associated with several classes overlap, Fuzzy ARTMAP tends to create about one category per data point. This phenomenon is termed *category proliferation*. It is a direct result of the fact that the classes are not separable and that Fuzzy ARTMAP will create a new category whenever the training point cannot be assigned to the correct class by match tracking. While this results in nearly perfect classification of the training vectors, performance on the test data (i.e. generalization) is often worse than if fewer categories had formed. Furthermore, the time required for the Fuzzy ART search and the memory required to store the weights are both proportional to the number of categories, providing additional incentives to limit the number of categories created.

Below, Section 2 reviews the original Fuzzy ARTMAP algorithm. The following section identifies specific objectives for the choice function that, if met, promote effective classification and learning. Section 4 introduces modifications to Fuzzy ARTMAP that ameliorate the category proliferation problem and lead to enhanced classification accuracy. Section 5 compares the performance of the standard and modified Fuzzy ARTMAP algorithms and Learning Vector Quantization (LVQ) [5] on actual data from the speech and image understanding domains. Finally, Section 6 summarizes the key concepts and results.

2 Standard Fuzzy ARTMAP Algorithm

2.1 Fuzzy ART Categorization

Fuzzy ART assigns vectors to categories based on two separate distance criteria, *match* and *choice*. For input vector $\boldsymbol{I}$ and category j, the *match function* is defined by

$$S_j(\boldsymbol{I}) \equiv \frac{|\boldsymbol{I} \wedge \boldsymbol{w}_j|}{|\boldsymbol{I}|},$$

where $\boldsymbol{w}_j$ is an analog-valued weight vector associated with category j, $\wedge$ denotes the fuzzy AND operator, $(\boldsymbol{p} \wedge \boldsymbol{q})_i \equiv min(p_i, q_i)$, and the norm $|\cdots|$ is defined by $|\boldsymbol{p}| \equiv \sum_i |p_i|$. The *Weber law choice function* is defined by

$$T_j(\boldsymbol{I}) \equiv \frac{|\boldsymbol{I} \wedge \boldsymbol{w}_j|}{\alpha + |\boldsymbol{w}_j|}, \tag{1}$$

where α is a small constant. Increasing α biases the search more toward categories with large $|\boldsymbol{w}_j|$. Fuzzy ART assigns each input vector to the category that maximizes $T_j(\boldsymbol{I})$ while satisfying $S_j(\boldsymbol{I}) \geq \rho$ (the *match criterion*), where the *vigilance*, ρ, is in the range $0 \leq \rho \leq 1$. *Unused categories* have match and choice values that are independent of $\boldsymbol{I}$, 1.0 and T_{unused}, respectively.

2.2 Aligning Categories with Classes via the MAP Field

The MAP field may be envisioned as a set of N_{cat} processing elements, each with N_{cla} weights, where N_{cat} is the number of categories formed by the Fuzzy ART module and N_{cla} is the total number of classes. It is basically used as an N_{cat} by N_{cla} look-up table. During training, if category J wins the Fuzzy ART competition and MAP field entry (J, $class_J$) of the table is not greater than some threshold, ρ^{ab}, ρ is increased and the Fuzzy ART search is repeated. If MAP field entry (J, $class_J$) is greater than ρ^{ab}, the data point is considered to have been correctly classified. Typically, one weight of each MAP field processing element equals 1.0 and the others equal 0, and $\rho^{ab} = 0$. This associates a single class with each category.

2.3 Updating the Weights

Once the training vector has been assigned to the correct class, the weights of the winning category are updated by

$$w_{Ji}^{new} = \begin{cases} w_{Ji}^{old} & w_{Ji} \leq I_i \\ w_{Ji}^{old} - \beta\left(w_{Ji}^{old} - I_i\right) & w_{Ji} > I_i \end{cases}, \tag{2}$$

where $0 < \beta \leq 1$.

If the training vector is assigned to an unused category, the weight vector of that category is set equal to the training vector and the associated MAP field weights are adjusted to reflect the training vector's class.

2.4 Complement Coding and Weight Rectangles

A precoding scheme, referred to as *complement coding*, is generally used [6], [7] with Fuzzy ART and is always used [4] with Fuzzy ARTMAP. Given M-dimensional feature vectors $\boldsymbol{a}$, $2M$-dimensional input vectors $\boldsymbol{I} = (\boldsymbol{a}, \boldsymbol{a}^c)$ are generated, where $a_i^c \equiv (1 - a_i)$.

The weight vectors are also $2M$-dimensional. Each weight, $\boldsymbol{w}_j$, may be envisioned as a pair of points, $\boldsymbol{u}_j$ and $\boldsymbol{v}_j^c$, in the M-dimensional feature space:

$$\boldsymbol{w}_j = (w_{j1}, w_{j2}, \ldots, w_{j2M}) = (\boldsymbol{u}_j, \boldsymbol{v}_j^c) = (u_{j1}, u_{j2}, \ldots, u_{jM}, v^c_{j1}, v^c_{j2}, \ldots, v^c_{jM}).$$

As shown in Figure 2, these points define a hyper-rectangle with corners $(\boldsymbol{u}_j, \boldsymbol{v}_j)$, where $v_{ji} = (v^c_{ji})^c = (1 - v^c_{ji})$. The *rectangle size* is defined as

$$\left|\boldsymbol{R}_j\right| \equiv \sum_{i=1}^{M} v_{ji} - u_{ji}.$$

As specified above, when an input vector is assigned to an uncommitted category, that weight vector is set equal to the input vector:

$$\boldsymbol{w}_j^{init} = (\boldsymbol{a}, \boldsymbol{a}^c) \Rightarrow (\boldsymbol{u}_j^{init}, \boldsymbol{v}_j^{init}) = (\boldsymbol{a}, (\boldsymbol{a}^c)^c) = (\boldsymbol{a}, \boldsymbol{a}) \Rightarrow \left|\boldsymbol{R}_j^{init}\right| = 0.$$

Thus, each newly initialized weight corresponds to a rectangle of zero size, a point in M-space.

The components of $w_j = (u_j, v_j^c)$ still decrease monotonically during learning. Since v_j^c moves monotonically toward the origin, v_j moves monotonically away from the origin. I.e., the size of the weight rectangle increases monotonically. As shown in Figure 2, v_j^c may be thought of as being plotted along axes that are inverted and offset from the axes along which u_j and v_j are plotted.

The norm of every input vector, $|I|$, equals M, the dimension of a. Consequently, the match function becomes $|I \wedge w_j| / M$, and the match criterion becomes $|I \wedge w_j| \geq \rho M$. Furthermore, $|I \wedge w_j| = M - |R_{aj}|$, where $|R_{aj}|$ is the size of the smallest rectangle enclosing u_j, v_j, and a. Thus, the match criterion specifies that a category can only capture an input vector if the size of the rectangle defined by the input and weight vectors is less than or equal to $M(1 - \rho)$.

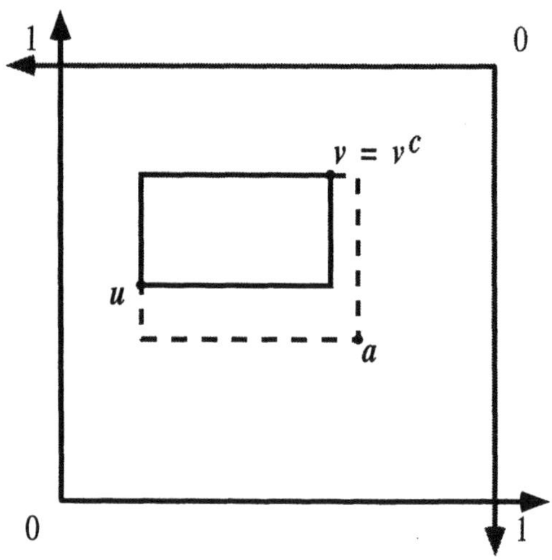

Figure 2. Geometric interpretation [6] of complement-coded input and weight vectors. Each $2M$-dimensional weight vector, w_j, may be thought of as two M-dimensional vectors, u_j and v_j, that define a hyper-rectangle in M dimensions. Each $2M$-dimensional input vector, I, may be thought of as two coinciding M-dimensional vectors at point a.

3 Desired Choice Function Characteristics

The choice function specifies how to measure the distance from a Fuzzy ART weight rectangle to an input vector in order to determine to which category to assign each input pattern. Since the size of the weight rectangle enters into the choice function, the distance measure is specific to each category – each category uses its own yardstick, and the yardsticks are not all the same.

Analyzing the choice function provides insight into Fuzzy ARTMAP learning. For example, immediately after the weights of a category are updated in response to an input vector, would that category capture an identical input vector?

The choice function should satisfy the following properties:

❶ Updating weight w_j in response to input vector $\boldsymbol{I}$ increases $T_j(\boldsymbol{I})$.

This is the basic premise behind learning. In Fuzzy ARTMAP, only the weights of a category with the same class association as the training vector are adjusted. This criterion establishes that the weight update facilitates the capture of an identical input pattern by that category in the future.

❷ An input vector that falls inside exactly one weight rectangle is captured by that category.

Each weight rectangle grows to become the smallest rectangle that encloses all vectors that have been captured by that category [6]. Consequently, associating a test vector with the class of the weight rectangle it lies in is a reasonable policy. Furthermore, this guarantees that a category created in response to some training vector that did not fall inside any weight rectangle captures that vector if it is immediately presented again.

❸ If an input vector lies inside both of two overlapping rectangles, the smaller rectangle (corresponding to the larger $|\boldsymbol{w}_j|$) has the larger choice value.

A large category box may surround several smaller weight rectangles, each with its own class association. All of the vectors falling within one of the smaller rectangles may belong to the same class, whereas the large rectangle encompasses several classes. Thus, the smaller category is more specific than the larger one, and an input pattern that lies within both should be assigned to the smaller one. This also guarantees that if a category is created in response to some training vector that fell inside a weight rectangle with an incorrect class association, the new category captures that vector if it is immediately presented again (unless another rectangle with zero area already exists at the same position, as may happen when the training set contains two identical vectors with different class associations).

The following are also desirable, but not necessary:

❹ The value of the choice function decreases monotonically with increasing distance of the input vector from the weight rectangle.

This is equivalent to specifying that if an input vector is outside two identical weight rectangles, the value of the choice function associated with the nearer one is larger. This condition states that, all other things being equal, we would like to associate a test vector with the class of the nearest training vector, as in the k nearest neighbors algorithms [8] for $k = 1$. However, though Fuzzy ART weight rectangles develop in response to training vectors, the actual training values are not stored. It is not possible to guarantee nearest neighbor behavior without storing the actual training patterns.

❺ An input vector that is equidistant from two weight rectangles is captured by the larger one.

Intuitively, a small weight rectangle results from a very localized set of training points, whereas a large weight rectangle results from a spread-out set of training points. If a test point lies equidistant from the two rectangles, it lies more standard deviations away from the smaller rectangle than from the larger one, and therefore should be assigned to the latter.

It is relatively straightforward to demonstrate that the original Weber law choice function (eq. 1) satisfies objectives ❶, ❸, and ❹, usually satisfies ❷, and does not satisfy criterion ❺. Section 4.5 provides an example of such a derivation (for an alternative choice function).

4 Modifications to Fuzzy ARTMAP

4.1 Pruning

If those categories capturing the fewest training vectors are periodically removed, the number of categories in use approaches a steady-state value. Categories that are not removed can grow to capture training vectors that were previously captured by the pruned categories. If categories capturing less than N_{prune} training vectors are periodically removed, the absolute maximum number of remaining categories is N_{prune} / N_{train}, where N_{train} is the total number of training data points. In practice, the number of categories formed is usually significantly smaller than this.

A simple way to keep track of how many training points are captured by each category is to change slightly how the MAP field weights are updated. Instead of setting a weight to *1.0* when a new category is formed, the weight is incremented by some constant, δ_{hit}, every time a training vector is assigned to that category. The size of the MAP field weight now indicates how many training vectors the associated category has captured. In order to prevent the MAP field weights from growing beyond bounds, a decay term is added. We implement this by periodically dividing all MAP field weights by 2.

4.2 Restricted Creation of New Categories

The rate of creation of new categories can be limited. For example, if the same data set is presented repeatedly during training, the number of categories that is created during each iteration may be limited to some small value, such as the total number of classes. This causes existing categories to grow as much as possible, generally reducing the total number of categories formed.

4.3 Bi-Directional Weight Movement

The Fuzzy ART learning law assures that weight rectangles only grow with time and never become smaller. Consequently, after enough iterations through the training data, every training point will lie either within or on the boundary of the weight rectangle of the category which captures it. This tends to create rectilinear win region boundaries throughout the feature space, which is often not desirable.

The following learning law causes the weight components to be updated in the same way as the original learning rule if the input vector component lies outside the range of the weight rectangle along that dimension, but causes both weight components to move toward the input vector when the input vector component falls inside the range of the weight rectangle along that dimension:

$$w_{Ji}^{new} = \begin{cases} w_{Ji}^{old} & w_{J\,comp} \geq I_{comp} \\ w_{Ji}^{old} - \beta\left(w_{Ji}^{old} - I_i\right) & w_{J\,comp} < I_{comp} \end{cases},$$

where $w_{J\,comp} = w_{J(i+M)}$ and $I_{comp} = I_{i+M}$ for $i \leq M$ and $w_{J\,comp} = w_{J(i\text{-}M)}$ and $I_{comp} = I_{i\text{-}M}$ for $i > M$.

The weight rectangle grows along the dimensions in which the input vector component lies outside the weight vector components and shrinks along the dimensions in which the input vector component lies between the weight vector components. In the former case, only one of the two weight components is updated, whereas in the latter case, both weight components are modified. This results in less rectilinear win regions that tend to better approximate the shapes of the optimal decision boundaries between intersecting class distributions.

Note that without pruning or restricted creation of new categories, bi-directional weight movement often produces a larger number of categories. Thus, one of the above methods for limiting the number of categories must generally be used for bi-directional weight movements to be beneficial.

Another learning law has previously been proposed [9] for bi-directional weight movement:

$$w_{Ji}^{new} = w_{Ji}^{old} - \beta\left(w_{Ji}^{old} - I_i\right),$$

regardless of the relationship between w_{Ji} and I_i. This update method always causes both components of the weight vector in each dimension of the feature space to move in proportion to their distance from the input vector component. Thus, the more distant weight component moves toward the input vector component *more* than the nearer weight component, and the rule guarantees that the weight rectangles only decrease in size. Since each weight rectangle is initialized as a single point in the M-dimensional feature space, this learning law causes the weight components to always move together, and, undesirably, all weight rectangles forever remain zero size. Markuzon found that her modified learning law yielded slightly increased classification accuracy at the expense of reduced code compression, i.e., the formation of a larger number of categories.

4.4 Category Re-Labeling

If the match function $S_J(\boldsymbol{I})$ is close to 1.0 (implying that the training point fell close to a very small weight rectangle), and yet the classes associated with category J and training vector $\boldsymbol{I}$ differ, standard Fuzzy ARTMAP creates a new category. Since tiling even a small volume of the feature space with very small rectangles requires a large number of categories, this is generally undesirable. Instead, it may be desirable to increment the MAP field entry $(J, class_I)$ if $S_J(\boldsymbol{I}) > \rho_{relabel}$, where $\rho_{relabel}$ is a threshold constant close to 1. This leads to two non-zero entries in the MAP field, associating one category with two classes. When a test vector is captured by such a category, it is possible to assign the vector a fuzzy membership in each of the classes, or to output a single class label based on the relative magnitudes of the two MAP field entries (consistent with Section 4.1).

Lim and Harrison [10] also suggested a modification to Fuzzy ARTMAP that facilitates the creation of one-to-many mappings between points in feature space and classes. The difference between the two approaches is that whereas category re-labeling associates one category with multiple classes, Lim and Harrison intentionally create multiple *tiny* Fuzzy ART categories (with different class associations) that are superimposed at the same position in the feature space. This improves the classification accuracy of one-dimensional input vectors. However, the number of fixed size weight rectangles required to tile a space is exponential in the input dimensionality. Consequently, Lim and Harrison found that their method provides little benefit in terms of classification accuracy for higher-dimensional spaces and aggravates the problem of category proliferation.

If category re-labeling is utilized, it is important to also choose the choice function favoring larger weight rectangles (Section 4.5) to reduce the likelihood that a category associated with multiple classes captures a point that can be unambiguously assigned to a single class.

4.5 Choice Function Favoring Larger Categories

As a result of the original Weber law choice function (eq. 1), if a data point falls inside both of two overlapping weight rectangles, the smaller rectangle (corresponding to the larger $|w_j|$) wins. This guarantees that a category created in response to some training vector captures the point if it is immediately presented again. Furthermore, if a data point falls midway between two weight rectangles, the smaller one wins. While it is necessary to preserve the former property, the latter one has undesirable consequences. In particular, a new category will capture data points from a large volume of feature space previously captured by a nearby large category.

Intuitively, a small weight rectangle results from a very localized set of training points, whereas a large weight rectangle results from a spread out set of training points. If a test point lies equidistant from the two rectangles, it lies more standard deviations away from the smaller rectangle than from the larger one, and therefore should be assigned to the latter.

The following alternative choice function provides the desired classification:

$$T_j^L(\boldsymbol{I}) \equiv |w_j|\left(1 + \left[|\boldsymbol{I} \wedge w_j| - |w_j|\right]/\alpha\right) \tag{3}$$

As in the Weber law choice function, α is a small constant. Here, α determines the range at which a larger weight rectangle is favored over a smaller one. $T_j^L(\boldsymbol{I})$ may be less than 0. Note that the quantity $|\boldsymbol{w}_j|$ need only be computed when the weights of category j are updated.

Figure 3 plots the decision boundaries between a small and a large weight rectangle in a two-dimensional feature space for the Weber law choice function and the alternative choice function favoring large weight rectangles. The remainder of this section derives that the latter choice function satisfies conditions ❶, ❸, and ❹ from Section 3 and usually achieves objectives ❷ and ❺.

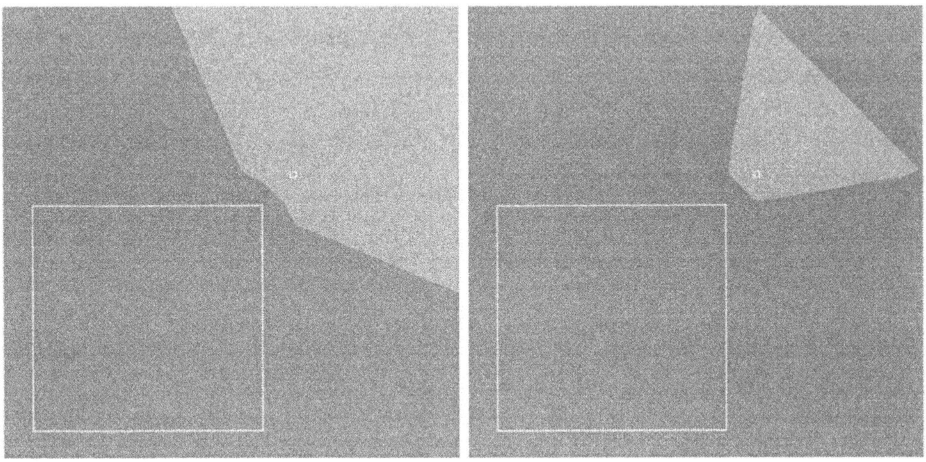

Figure 3. Win regions of two categories, one with a large weight rectangle and one with a small weight rectangle, when the Weber law choice function is used (left) and when the alternative choice function favoring larger weight rectangles is used (right).

❶ Does enlarging weight rectangle j toward input vector $\boldsymbol{I}$ increase $T_j^L(\boldsymbol{I})$?

Consider an input vector, $\boldsymbol{I}$, that lies outside weight rectangle $\boldsymbol{R}_j$. This implies that some component, i, of the input vector is smaller than the corresponding weight: $I_i < w_{ji}$.

$$\frac{\partial T_j^L(\boldsymbol{I})}{\partial w_{ji}} = \frac{\partial |\mathbf{w}_j|\left(1+\left[|\boldsymbol{I} \wedge \mathbf{w}_j| - |\mathbf{w}_j|\right]/\alpha\right)}{\partial w_{ji}}$$

$$= \frac{\partial \left(w_{ji} + |\mathbf{w}'_j|\right)\left(1+\left[I_i - w_{ji} + |\boldsymbol{I}' \wedge \mathbf{w}'_j| - |\mathbf{w}'_j|\right]/\alpha\right)}{\partial w_{ji}}$$

$$= 1 - |\mathbf{w}_j|/\alpha + \left[|\boldsymbol{I} \wedge \mathbf{w}_j| - |\mathbf{w}_j|\right]/\alpha.$$

Note that $|I \wedge w_j| < |w_j|$. Furthermore, the match criterion guarantees that $|w_j| \geq \rho M$. Choosing $\rho \geq \alpha/M$ assures that $|w_j|/\alpha \geq 1$. Since α is normally a very small number, this requirement does not impose serious restrictions on the choice of ρ. If α and ρ obey this relationship, the derivative is negative, and decreasing w_{ji}, i.e., moving the weight toward the training vector, increases $T_j^L(I)$. ✓

❷ <u>If *I* falls inside exactly one weight rectangle, is it captured by that category?</u>

Assume that I lies inside R_j but not in R_k. This implies that $|I \wedge w_j| = |w_j|$. Consequently,

$$T_j^L(I_{inside}) = |w_j|\left(1 + \left[|I_{inside} \wedge w_j| - |w_j|\right]/\alpha\right) = |w_j| . \tag{4}$$

Define $\delta = |w_k| - |I \wedge w_k|$. I.e., δ is the L_1 distance of I from weight rectangle k. Hence,

$$T_k^L(I) = |w_j|\left(1 + \left[|I \wedge w_j| - |w_j|\right]/\alpha\right) = |w_k|(1 - \delta/\alpha) .$$

I is captured by category j if and only if

$$T_j^L(I_{inside}) > T_k^L(I) \quad \Leftrightarrow \quad |w_j| > |w_k|(1 - \delta/\alpha) \quad \Leftrightarrow \quad \delta > \frac{\alpha\left(|w_k| - |w_j|\right)}{|w_k|} .$$

This is nearly identical to the condition for the Weber law choice function. A small rectangle (large $|w_k|$) will capture an input vector falling near its boundary (distance δ) and falling inside a larger rectangle R_j. If I is presented twice in a row, Fuzzy ARTMAP with fast learning and at least one uncommitted unit will classify it correctly the second time.

❸ <u>If *I* lies inside both of two overlapping rectangles, does the smaller rectangle win?</u>

From eq. 4, if the input vector lies inside R_j, $T_j^L = |w_j|$. Since the smaller rectangle has the larger $|w|$, it captures the input vector. ✓

❹ <u>Does moving $\boldsymbol{I}$ away from weight rectangle j decrease $T_j^L(\boldsymbol{I})$?</u>

Consider an input vector, $\boldsymbol{I}$, that lies outside weight rectangle $\boldsymbol{R}_j$. This implies that some component, i, of the input vector is smaller than the corresponding weight: $I_i < w_{ji}$.

$$\frac{\partial T_j^L(\boldsymbol{I})}{\partial I_i} = \frac{\partial |\boldsymbol{w}_j|\left(1+\left[|\boldsymbol{I}\wedge\boldsymbol{w}_j|-|\boldsymbol{w}_j|\right]/\alpha\right)}{\partial I_i} = \frac{|\boldsymbol{w}_j|}{\alpha}.$$

Since the derivative is positive, decreasing I_i, i.e., moving the input vector away from the weight rectangle, decreases $T_j^L(\boldsymbol{I})$. ✓

❺ <u>When $\boldsymbol{I}$ is not in any weight rectangle, which is favored, small or large rectangles?</u>

Consider an input vector, $\boldsymbol{I}$, that lies outside weight rectangles $\boldsymbol{R}_j$ and $\boldsymbol{R}_k$. Define $\delta_j = |\boldsymbol{w}_j| - |\boldsymbol{I}\wedge\boldsymbol{w}_j|$ and $\delta_k = |\boldsymbol{w}_k| - |\boldsymbol{I}\wedge\boldsymbol{w}_k|$. These are the L_1 distances of $\boldsymbol{I}$ from weight rectangles j and k, respectively. If $T_j^L = T_k^L$, then

$$|\boldsymbol{w}_j|(1-\delta_j/\alpha) = |\boldsymbol{w}_k|(1-\delta_k/\alpha) \;\Rightarrow\; \delta_j = \frac{|\boldsymbol{w}_k|}{|\boldsymbol{w}_j|}\delta_k + \alpha\left(1-\frac{|\boldsymbol{w}_k|}{|\boldsymbol{w}_j|}\right) \equiv C\delta_k + D.$$

The range, δ_j, at which the choice function of category j equals the choice function of category k is a linear function of the distance, δ_k, from $\boldsymbol{I}$ to $\boldsymbol{R}_k$. The slope (C) and the ordinate offset (D) of the line are functions of the sizes of the weight rectangles. If $\boldsymbol{R}_j$ is smaller than $\boldsymbol{R}_k$ then $|\boldsymbol{w}_j| > |\boldsymbol{w}_k|$, $C < 1$, and $D > 0$. Thus, the first term favors large rectangles and the second term favors small rectangles. The second term dominates only when $\boldsymbol{I}$ is very close to the two weight rectangles ($\delta_k \cong 0$). In order to reduce the range in which small rectangles are favored, α is chosen as small as possible. Thus, if $\boldsymbol{I}$ is equidistant from two weight rectangles, it will be captured by the larger one unless it lies within a minuscule distance of both $\boldsymbol{R}_j$ and $\boldsymbol{R}_k$.

4.6 Multiplication-Free Choice Function

The Weber law choice function and the choice function favoring larger categories are represented in mathematically compact form in equations 1 and 3. However, neither strictly satisfies all of the choice function objectives identified in Section 3. The following function does:

$$T_j^M(\boldsymbol{I}) \equiv \begin{cases} |\boldsymbol{w}_j| & \boldsymbol{I} \text{ inside } \boldsymbol{R}_j \\ |\boldsymbol{I} \wedge \boldsymbol{w}_j| - (1+\alpha)|\boldsymbol{w}_j| & \text{else.} \end{cases} \tag{5}$$

The characteristic that input vector $\boldsymbol{I}$ lies inside weight rectangle $\boldsymbol{R}_j$ if and only if $|\boldsymbol{I} \wedge \boldsymbol{w}_j| = |\boldsymbol{w}_j|$ is utilized to determine which branch of the condition to evaluate for any particular input vector and category.

As illustrated in Figure 4, if $\alpha = 0$, input vectors that do not lie inside any weight rectangle are captured by the nearest rectangle, regardless of size. Larger values of α provide an additive bias favoring larger weight rectangles.

Figure 4. Win regions of two categories based on the multiplication-free choice function with $\alpha = 0$. Input vectors are captured by the closest (according to Manhattan distance) weight rectangle, regardless of the sizes of the rectangles.

One attractive feature of Fuzzy ART for hardware implementation is that it relies largely on very simple operations. Instead of the usual multiplication, a simpler comparison is carried out on each corresponding pair of input and weight vector components. Furthermore, the multiplication by β in the learning rule (eq. 2) may be implemented as a shift (if β is a power of 2) or replaced by a subtraction. Thus, choosing $\alpha = 0$ (no multiplication required) or $\alpha = 1$ (multiply by shifting) in equation 5 makes it possible to implement the entire learning algorithm without multipliers and facilitates extremely dense hardware implementations.

Whereas the choice function favoring larger categories does not satisfy two of the criteria in (small) portions of the feature space, the multiplication-free choice function strictly satisfies all of the choice function objectives. Nonetheless, experimentally, the former function tends to provide somewhat better classification accuracy. Apparently, the more rectilinear decision boundaries resulting from the multiplication-free choice function are a drawback. Thus, while the multiplication-free choice function is potentially advantageous for custom hardware implementations, it is not recommended when running on a general-purpose computer (which invariably includes fast multipliers).

4.7 Anisotropic Distance Measure

A common issue with learning algorithms is how to pre-scale the feature vector components. If a great deal of variation of one component has no effect on the classification of inputs, but a small variation in another component changes the classification, the resulting performance tends to be worse than if the roles are reversed. Re-scaling the feature space may significantly change the classification accuracy. The anisotropic choice function presented here assigns more significance to variation in those dimensions in which less intra-class variation has been observed, as indicated by the lengths of those sides of the weight rectangles. This may be viewed as *locally* re-scaling the feature space in order to represent elongated class distributions.

With complement coding, Fuzzy ART utilizes *two* weights per input vector dimension per category, twice as many as some other vector quantization algorithms. With *k*-means methods [11] and LVQ [5], each set of weights only indicates the position of a category. The Fuzzy ART weights explicitly provide additional information about the *extent* of the category's win region. This knowledge is used by the choice function to favor the smallest category when an input vector falls inside multiple weight rectangles. It is also used to limit the maximum extent of categories (via the vigilance criterion), and is instrumental to the Fuzzy ARTMAP supervised learning mechanism, which restricts attention to more specific (smaller) categories after an input is initially associated with an incorrect class.

Whereas knowledge about the size of the weight rectangles is utilized by Fuzzy ART, knowledge about their shapes (aspect ratios), though also available, is not. The following choice function makes use of this information to provide a degree of local auto-scaling:

$$T_j^A(\boldsymbol{I}) \equiv \begin{cases} |\boldsymbol{w}_j| & \boldsymbol{I} \text{ inside } \boldsymbol{R}_j \\ \left|\dfrac{(\boldsymbol{I} \wedge \boldsymbol{w}_j) - \boldsymbol{w}_j}{1+\alpha-\boldsymbol{w}_j-\boldsymbol{w}_{jc}}\right| = \sum\limits_i \dfrac{min(I_i \wedge w_{ji}) - w_{ji}}{1+\alpha - w_{ji} - w_{jic}} = -\sum\limits_i \dfrac{\delta_{ji}}{\alpha + R_{ji}} & \text{else.} \end{cases} \tag{6}$$

Here, α is a small constant. If w_{ji} is the weight corresponding to input vector component I_i, w_{jic} is the complementary weight corresponding to I_i^c. Using the notation of complement coding (Section 2.4), for $i < M$, $w_{ji} = u_{ji}$ and $w_{jic} = v^c{}_{ji}$. For $i \geq M$, the roles are reversed. δ_{ji} is the distance of the input vector from weight rectangle j in dimension i. Either δ_{ji} or $\delta_{j(i+M)}$ will be zero ($i < M$). If $\boldsymbol{I}$ lies inside $\boldsymbol{R}_j$ in dimension i, both δ_{ji} and δ_{jic} are zero. R_{ij} is the length of $\boldsymbol{R}_j$ in dimension i:

$$R_{ji} \equiv v_{ji} - u_{ji} = \left(1 - v_{ji}^c\right) - u_{ji} = 1 - u_{ji} - v_{ji}^c \,.$$

In each dimension, the contribution of the distance from $\boldsymbol{I}$ to $\boldsymbol{R}_j$ is scaled by α plus the side length of the rectangle. The constant α is necessary to prevent division by zero. Also, increasing α decreases how much small rectangle dimensions are penalized relative to large ones.

Figure 5. Win regions of two symmetrical categories based on the anisotropic choice function with $\alpha = 0.01$ (left) and $\alpha = 0.2$ (right).

Figure 5 shows the win regions of two symmetrical categories based on the anisotropic choice function with $\alpha = 0.01$ and $\alpha = 0.2$. Large weight rectangles are favored in the competition for points not falling inside any weight rectangle. Figure 6 compares the win regions of three elongated weight rectangles based on the Weber law, the choice function favoring larger weight rectangles, the multiplication-free choice function, and anisotropic choice function.

The anisotropic distance measure satisfies all five of the choice function objectives defined in Section 3. Furthermore, experimentally, this measure provided slightly improved classification accuracy in some instances.

5 Results

The ultimate validation of any algorithm is its performance on real problems. This section compares the performance of the standard and modified Fuzzy ARTMAP algorithms and Optimized Learning Rate Learning Vector Quantization [5] on actual data from the speech and image understanding domains. The tasks include varying degrees of noise and intersection of the class distributions. The dimensionality of the input vectors ranges from 2 to 12. Overall, the modified algorithm out-performs the original Fuzzy ARTMAP by a wide margin and provides comparable performance to that of OLVQ.

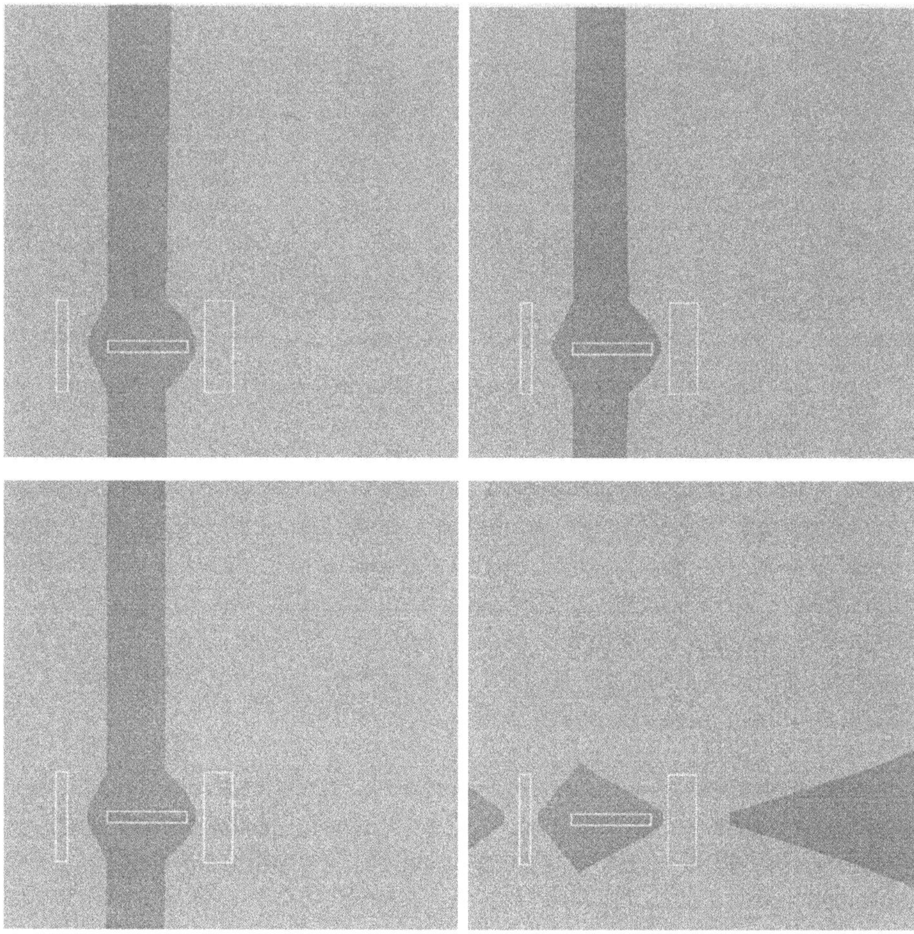

Figure 6. Win regions of three categories with elongated weight rectangles based on the Weber law (top left), the choice function favoring larger weight rectangles (top right), the multiplication-free choice function with $\alpha = 0$ (bottom left), and the anisotropic choice function with $\alpha = 0.01$ (bottom right). The two vertically oriented rectangles are associated with the same class. Since there is little difference in the sizes of the weight rectangles, there is little difference between the win regions of the first three choice functions.

5.1 Speaker-Independent Vowel Classification

The Peterson-Barney vowel data [3] consists of 1520 data vectors resulting from two utterances of each of the ten American vowels by each of 76 speakers. Thus, there are ten classes in the data set. Each four-dimensional vector consists of the first four formant frequencies. The data was randomly separated into a training set of 1020 points and a test set consisting of the remaining 500 points. The task is to learn a classifier on the training data and use it to assign each test vector to the appropriate class.

Considering only the first and second formant frequencies, several sets of identical training vectors that are labeled with multiple classes occur. Consequently, it is possible for the number of categories generated by the standard algorithm to grow without bound. In fact, the standard Fuzzy ARTMAP parameter set yielding the best classification (71% correct) generated 620 categories after 20 iterations through this training set. The smallest number of categories generated by the standard algorithm after one pass through the training data was 104, giving 57% correct classification of the test set. In contrast, the modified algorithm achieved up to 72% correct classification with 105 categories and 74% correct with 141 categories. Table 1 and Figure 7 illustrate the performance of the Fuzzy ARTMAP variants on the Peterson-Barney vowel data in two dimensions.

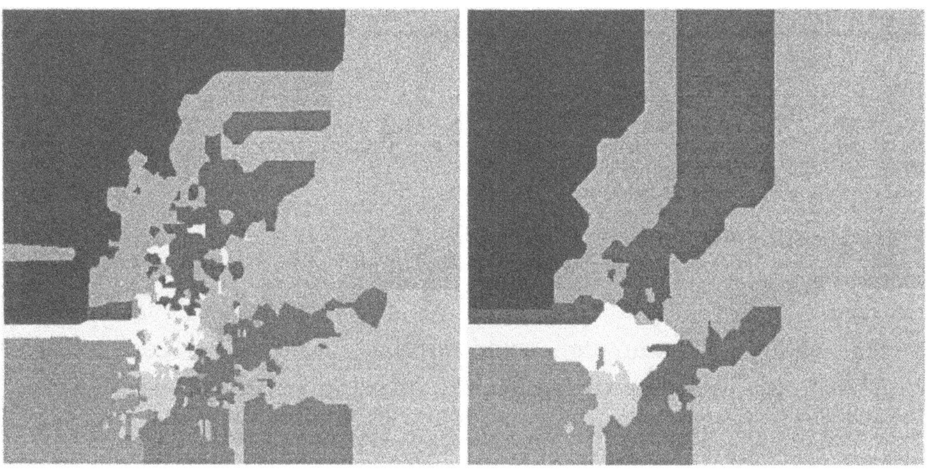

Figure 7. Map of the feature space generated by the standard Fuzzy ARTMAP algorithm, left, and by the modified algorithm using bi-directional weight movement, limited category creation, pruning, and the alternative choice function, right. The numerous small categories in the left plot indicate over-training by the standard algorithm. Both variants were trained on 1020 points randomly chosen from the Peterson-Barney vowel data plotted in Figure 1.

Table 1. Performance of Fuzzy ARTMAP on the Peterson-Barney vowel data (formant frequencies f_1 and f_2). The parameter values shown here yielded the highest performance. Results of numerous simulations yielding equivalent or worse performance are not included. The **bold** parameters were used to generate the plots in Figure 7.

α	β	N_{iters}	Max. cats per iteration	N_{prune}	ρ	$\rho_{relabel}$	Choice	Bi-directional weight movement	N_{cats}	% correct
Modified Fuzzy ARTMAP:										
.0001	**0.1**	**20**	**20**	**3**	**0.5**	-	**Alt.**	√	**141**	**74**
.0001	0.01	5	-	-	0.5	0.99	Alt.		121	73
0.01	0.05	5	-	-	0.5	0.99	Alt.		96	72
0.01	0.05	5	-	-	0.5	0.99	Weber		105	72
0.01	0.05	20	-	3	0.5	-	Weber		260	72
.0001	0.05	40	20	5	0.5	-	Alt.	√	109	71
0.01	0.1	5	-	-	0.5	0.99	Alt.		87	69
Standard Fuzzy ARTMAP:										
0.01	**0.05**	**20**	-	-	**0.5**	-	**Weber**		**620**	**71**
0.01	0.05	5	-	-	0.5	-	Weber		450	69
0.01	1.0	5	-	-	0.5	-	Weber		243	68
0.01	1.0	2	-	-	0.5	-	Weber		181	66
0.01	0.25	5	-	-	0.5	-	Weber		247	65
0.01	1.0	1	-	-	0.5	-	Weber		104	57

Note that this is a very difficult classification task – for comparison, the best classification accuracy achieved with Optimized Learning Rate LVQ1 [12] initialized by *k*-means clustering was 70.2% correct with 100 codebook vectors.

As is evident from Table 1, numerous combinations of the modified Fuzzy ARTMAP parameters yield similar classification performance. This task identified a particular set of parameters that is relatively effective for a wide variety of data sets. To demonstrate the robustness of the modified algorithm, all of the subsequent experiments are based on this same parameter set. It would be possible to achieve somewhat better classification accuracy and/or smaller numbers of categories by fine-tuning the parameters for each data set.

Table 2. Classification of four-dimensional formant frequency vectors. Each value is the average from five training sessions with identical data sets and parameters, but different randomly selected orderings of the training points. For all algorithms, 16,000 training vector presentations occurred. The standard Fuzzy ARTMAP parameters were $\alpha = 0.0001$, $\beta = 1.0$, and $\rho_0 = 0.5$. The modified algorithm was run with $\alpha = 0.0001$, $\beta = 0.1$, $\rho_0 = 0.5$, pruning of categories listing fewer than *3* vectors captured during the last 300 vector presentations, at most *20* initial categories, addition of at most one unused category every 250 vector presentations, the choice function favoring larger categories, and bi-directional weight movement.

Algorithm	Average N_{cats}	Average Accuracy
Standard Fuzzy ARTMAP	97	75.9%
Modified Fuzzy ARTMAP	69	79.5%
OLVQ	70	70.8%

The performances of the three algorithms on the task of classifying vectors composed of all four formant frequencies are similar to the performances on the two-dimensional data listed above. As detailed in Table 2, the standard Fuzzy ARTMAP algorithm yields a larger number of categories and lower classification accuracy than the modified algorithm. LVQ yields somewhat lower accuracy regardless of the number of codebook vectors selected.

5.2 Image Annotation

Another domain of great interest is image understanding. Below, the specific problem of *image annotation* [13] is considered. This consists of forming a vector per image pixel and classifying it as belonging to one of several subjects, in these examples "Field", "Forest", "Other", "Road", "Sky", and "Water". The "Other" class subsumes man-made structures such as bridges, buildings, cars, signs, and so on. Each data point is a twelve-dimensional vector whose components represent local color information and Haar wavelet transform coefficients [14].

The first task is to learn a classifier on training data from one side of an image and use it to assign each test vector from the other side of the same image to the appropriate class. The class and training/test set memberships boundaries are defined by closed contours in Figure 8.

Figure 8. Training and test data for the pixel vector classification task. The three boundaries drawn on the left side of the image define the training data for the classes "Sky", "Forest", and "Water". The boundaries on the right side of the image define the test data. The training and test data sets are non-overlapping.

The performances of the three algorithms on the pixel vector classification task are similar to the performances on the vowel classification tasks above. As detailed in Table 3, the standard Fuzzy ARTMAP algorithm yields a larger number of categories and lower classification accuracy than the modified algorithm. LVQ yields accuracy comparable to that of the modified Fuzzy ARTMAP regardless of the number of codebook vectors selected. The recognition rates of all three algorithms are higher and the number of categories are lower than in the vowel classification task because the class distributions overlap less in this data set.

The second task is to learn a classifier on training data from six images of diverse outdoor scenes and use it to assign each test vector from yet another distinct scene to the appropriate class. The class memberships in the training set are defined by closed contours as in Figure 8. Performance on the test set is assessed visually via Figure 9 rather than numerically.

Due to the diverse images used for the training set, the class distributions underlying this data set intersect significantly. For the same reason, the test image provides a good assessment of the generalization abilities of the algorithms. As in the other tasks, the standard Fuzzy ARTMAP algorithm over-trained (612 categories) and, as evidenced by the mottled appearance of the annotated test image, provided poor generalization. The modified Fuzzy ARTMAP did better, producing 76 categories and improved classification of the test data. Of the three algorithms, OLVQ yielded the most accurate and uniform classification.

Table 3. Classification of twelve-dimensional pixel vectors. Each value is the average from five training sessions with identical data sets and parameters, but different randomly selected orderings of the training points. For all algorithms, 40,000 training vector presentations occurred. The standard Fuzzy ARTMAP parameters were $\alpha = 0.0001$, $\beta = 1.0$, and $\rho_0 = 0.5$. The modified algorithm was run with $\alpha = 0.0001$, $\beta = 0.1$, $\rho_0 = 0.5$, pruning of categories listing fewer than *3* vectors captured during the last 300 vector presentations, at most *20* initial categories, addition of at most one unused category every 250 vector presentations, the choice function favoring larger categories, and bi-directional weight movement.

Algorithm	Average N_{cats}	Average Accuracy
Standard Fuzzy ARTMAP	12.2	98.72%
Modified Fuzzy ARTMAP	9.4	99.53%
OLVQ	10	99.71%

6 Conclusions

Due to category proliferation, the standard Fuzzy ARTMAP algorithm typically over-trains and provides poor generalization on data drawn from intersecting class distributions. This problem can be ameliorated by pruning categories, restricting the creation of categories, bi-directional weight movement, category re-labeling, and using a choice function that favors larger categories when the data point is outside the categories' weight rectangles. The improved performance is clearly evident on tests with real-world data sets of varying dimensionality and varying degrees of overlap among the class distributions. The learning parameters were *not* optimized for each data set. Nonetheless, the modified Fuzzy ARTMAP algorithm performed comparably and in some cases better than Optimized Learning Rate LVQ.

Figure 9. Test image (upper left) and annotated images produced by Optimized Learning Rate LVQ (upper right), standard Fuzzy ARTMAP (lower left), and modified Fuzzy ARTMAP (lower right) algorithms. In the annotated images, the gray values indicate "Road", "Water" (only a few tiny regions), "Forest", "Sky", "Field", and "Other", in order of darkest to lightest. For all algorithms, 100,000 training vector presentations occurred. OLVQ was configured to produce 200 codebook vectors. The standard Fuzzy ARTMAP parameters were $\alpha = 0.0001$, $\beta = 1.0$, and $\rho_0 = 0.5$. The modified algorithm was run with $\alpha = 0.0001$, $\beta = 0.1$, $\rho_0 = 0.5$, pruning of categories listing fewer than *3* vectors captured during the last 300 vector presentations, at most *20* initial categories, addition of at most one unused category every 250 vector presentations, the choice function favoring larger categories, and bi-directional weight movement.

Acknowledgments

The authors developed the algorithms described here while in the Department of Electrical and Computer Engineering at the University of California, San Diego. This research was supported by the Advanced Research Projects Agency (ARPA F49620-92-J-0520).

References

[1] Kohonen, T., Barna, G. and Chrisley, R. (1988), "Statistical pattern recognition with neural networks: benchmarking studies," *Proceedings of the IEEE International Conference on Neural Networks*, pp. 61-68.

[2] Bezdek, J. (1981), *Pattern Recognition with Fuzzy Objective Function Algorithms*, Plenum Press, New York.

[3] Peterson, G.E. and Barney, H.L. (1952), "Control methods used in a study of the vowels," *Journal of the Acoustical Society of America*, Vol. 24, pp. 175-184.

[4] Carpenter; G.A., Grossberg, S., Markuzon, N., Reynolds, J.H., and Rosen, D.B. (1992), "Fuzzy ARTMAP: a neural network architecture for incremental supervised learning of analog multidimensional maps," *IEEE Transactions on Neural Networks*, Vol. 3, pp. 698-713.

[5] Kohonen, T. (1990), "The self-organizing map," *Proceedings of the IEEE*, Vol. 78, pp. 1464-1480.

[6] Carpenter, G.A., Grossberg, S., and Rosen, D.B. (1991), "Fuzzy ART: fast stable learning and categorization of analog patterns by an adaptive resonance system," *Neural Networks*, Vol. 4, pp. 759-771.

[7] Huang, J., Georgiopoulos, M., and Heileman, G.L. (1995), "Fuzzy ART properties," *Neural Networks*, Vol. 8, pp. 203-213.

[8] Dasarathy, B.V. (1991), *Nearest Neighbor (NN) Norms: NN Pattern Classification Techniques*, IEEE Computer Society Press, Los Alamitos.

[9] Markuzon, N. (1994), "Handwritten digit recognition using fuzzy ARTMAP network," *Proceedings of the World Congress on Neural Networks*, pp. 117-122.

[10] Lim, C.P., and Harrison, R.F. (1997), "Modified fuzzy ARTMAP approaches Bayes optimal classification rates: an empirical demonstration," *Neural Networks*, Vol. 10, pp. 755-774.

[11] Ripley, B.D. (1996), *Pattern Recognition and Neural Networks*, Cambridge University Press, Cambridge.

[12] Kohonen, T., Kangas, J., Laaksonen, J., and Torkkola, K. (1992), "LVQPAK: A software package for the correct application of learning vector quantization algorithms," *Proceedings of the IJCNN International Joint Conference on Neural Networks*, pp. 725-730.

[13] Ma, W.Y. and Manjunath, B.S. (1995), "A comparison of wavelet transform features for texture image annotation," *Proceedings of the International Conference on Image Processing*, pp. 256-259.

[14] Blume, M. and Ballard, D. (1997), "Image annotation based on learning vector quantization and localized Haar wavelet transform features," *Proceedings of the SPIE*, Vol. 3077, pp. 181-190.

CHAPTER 4

SUPERVISED ADAPTIVE RESONANCE THEORY AND RULES

A.-H. Tan
Kent Ridge Digital Labs
National University of Singapore
Singapore

Supervised Adaptive Resonance Theory is a family of neural networks that performs incremental supervised learning of recognition categories (pattern classes) and multidimensional maps of both binary and analog patterns. This chapter highlights that the supervised ART architecture is compatible with IF-THEN rule-based symbolic representation. Specifically, the knowledge learned by a supervised ART system can be readily translated into rules for interpretation. Similarly, *a priori* domain knowledge in the form of IF-THEN rules can be converted into a supervised ART architecture. Not only does initializing networks with prior knowledge improve predictive accuracy and learning efficiency, the inserted symbolic knowledge can also be refined and enhanced by the supervised ART learning algorithm. By preserving symbolic rule form during learning, the rules extracted from a supervised ART system can be compared directly with the originally inserted rules.

1 Introduction

Supervised Adaptive Resonance Theory is an extension of Adaptive Resonance Theory (ART) to perform incremental supervised learning of recognition categories (pattern classes) and multidimensional maps of both binary and analog patterns. Two classical examples of supervised ART systems are ARTMAP [3], [4] and its bidirectional compressed variant, known as the Adaptive Resonance Associative Map (ARAM) [15]. Although both ARTMAP and ARAM has architectures that are compatible with rule-based representation, to facilitate discussion, I have used ARTMAP (in particular fuzzy ARTMAP) to illustrate the linkages between supervised ART systems and symbolic knowledge processing.

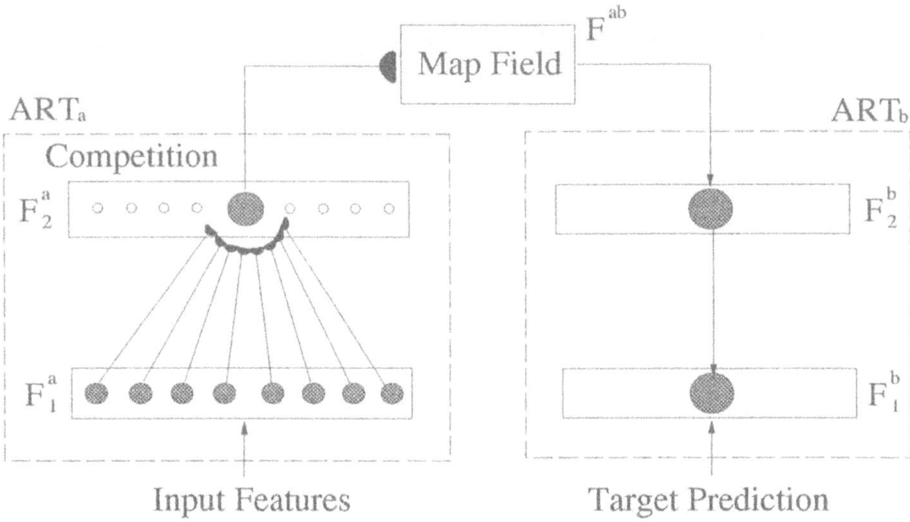

Figure 1. Rules in ARTMAP. Each node in the F_2^a field represents a recognition category of ART_a input patterns. Through the inter-ART map field, each such node is associated to an ART_b category in the F_2^b field, which in turn encodes a prediction. Learned weight vectors, one for each F_2^a node, constitute a set of rules that link antecedents to consequents. The number of rules equals the number of F_2^a nodes that become active during learning.

When performing classification tasks, ARTMAP formulates recognition categories of input patterns, and associates each category with its respective prediction. The knowledge that ARTMAP discovers during learning, is compatible with IF-THEN rule-based representation. Specifically, the recognition categories learned by the F_2^a category nodes are compatible with rules that link antecedents to consequents (Figure 1). At any point during the incremental learning process, the system architecture can be translated into a compact set of rules analyzable by human experts [6].

On the other hand, rules can be readily inserted into an ARTMAP network that can then be refined by learning from examples. To handle intermediate attributes, Cascade ARTMAP, a generalization of ARTMAP, represents rule cascades of rule-based knowledge explicitly [16]. During ARTMAP learning, new recognition categories (rules) are created dynamically to cover the deficiency of the domain theory. By the self-stabilizing property, learning in Cascade ARTMAP does not wash away existing knowledge and the meanings of units do not shift. This allows preservation of symbolic rules during neural network learning. Using a generalized ARTMAP rule extraction procedure, the final system states can be converted back to a compact set of rules. This enables direct comparison of the original knowledge with the refined rules. Also, each extracted rule is associated with a confidence factor that indicates its importance or usefulness. This allows ranking and evaluation of the extracted knowledge. In all, the Cascade ARTMAP rule insertion, refinement, and extraction procedures form a paradigm for symbolic knowledge refinement and evaluation (Figure 2).

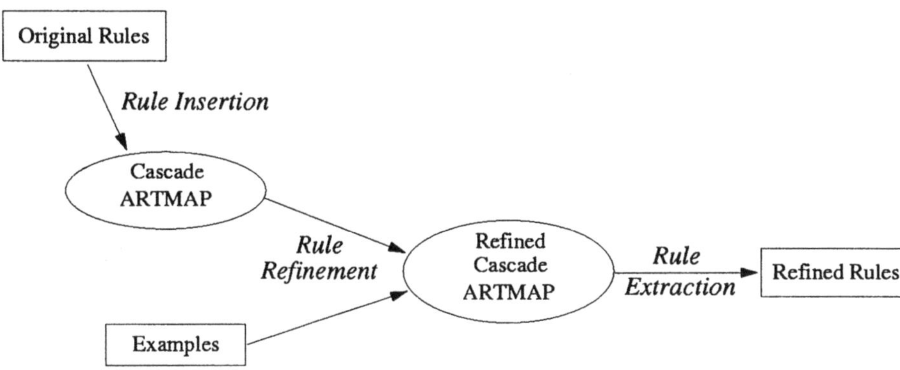

Figure 2. Cascade ARTMAP for symbolic knowledge refinement and evaluation.

The remaining sections of this chapter are organized as follows. To make this chapter self-contained, section 2 provides a summary of fuzzy ARTMAP. Section 3 motivates the generalization of fuzzy ARTMAP to Cascade ARTMAP and presents the Cascade ARTMAP rule insertion, rule chaining, rule refinement, and rule extraction algorithms. Section 4 illustrates Cascade ARTMAP's performance on a DNA promoter recognition problem. The final section states concluding remarks and highlights future applications.

2 Fuzzy ARTMAP

Fuzzy ARTMAP [3], a generalization of binary ARTMAP [4], learns to classify inputs by a pattern of fuzzy membership values between 0 and 1 indicating the extent to which each feature is present. This generalization is accomplished by replacing the ART 1 modules [2] of the binary

ARTMAP system with fuzzy ART modules [5]. Each ARTMAP system includes a pair of Adaptive Resonance Theory modules (ART_a and ART_b) that create stable recognition categories in response to arbitrary sequences of input patterns (Figure 3). An associative learning network and an internal controller link these modules to make the ARTMAP system operate in real time.

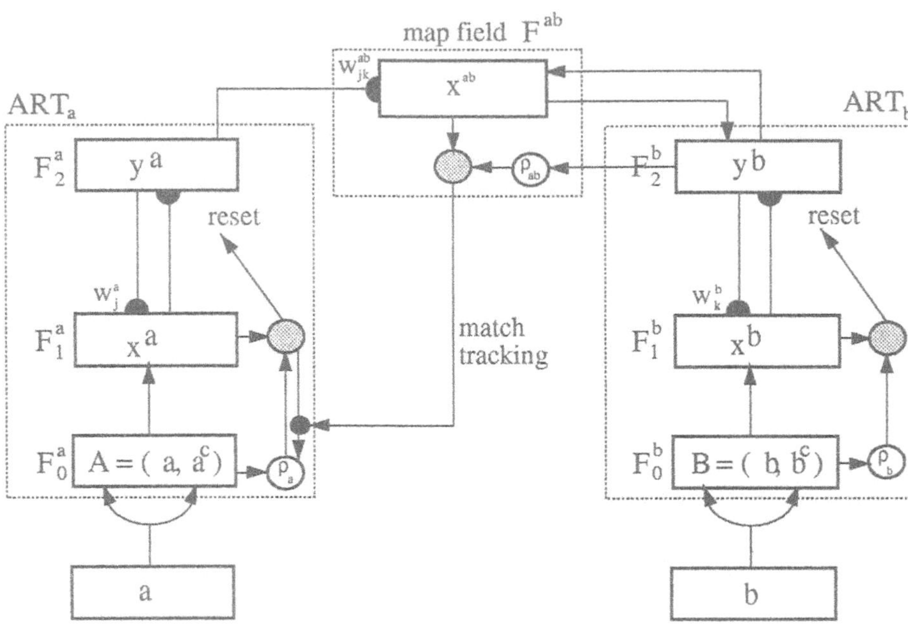

Figure 3. Fuzzy ARTMAP architecture. The ART_a complement coding preprocessor transforms the M_a-vector **a** into the $2M_a$-vector $\mathbf{A} = (\mathbf{a}, \mathbf{a}^c)$ at the ART_a field F_0^a. **A** is the input vector to the ART_a field F_1^a. Similarly, the input to F_1^b is the $2M_b$-vector $(\mathbf{b}, \mathbf{b}^c)$. When ART_b disconfirms a prediction of ART_a, map field inhibition induces the match tracking process. Match tracking raises the ART_a vigilance (ρ_a) to just above the F_1^a-to-F_0^a match ratio $|\mathbf{x}^a|/|\mathbf{A}|$. This triggers an ART_a search which leads to activation of either an ART_a category that correctly predicts **b** or to a previously uncommitted ART_a category node.

2.1 Fuzzy ART

Fuzzy ART [5] incorporates computations from fuzzy set theory into ART systems. By replacing the crisp (nonfuzzy) intersection operator ($\cap$) that describes ART 1 dynamics [2] by the fuzzy AND operator ($\wedge$) of fuzzy set theory, fuzzy ART can learn stable categories in response to either analog or binary patterns.

ART field activity vectors: Each ART system includes a field F_0 of nodes that represent a current input vector; a field F_1 that receives both bottom-up input from F_0 and top-down input from a field F_2 that represents the active code, or category. Vector $\mathbf{I}$ denotes F_0 activity; vector $\mathbf{x}$ denotes F_1 activity; and vector $\mathbf{y}$ denotes F_2 activity.

Weight vector: Associated with each F_2 category node j is a vector $\mathbf{w}_j$ of adaptive weights, or long-term memory (LTM) traces. Initially

$$w_{j1}(0) = \ldots = w_{jM}(0) = 1; \tag{1}$$

then each category is *uncommitted.* After a category codes its first input, it becomes *committed.*

Parameters: A choice parameter $\alpha > 0$, a learning rate parameter $\beta \in [0,1]$, and a vigilance parameter $\rho \in [0,1]$ determine fuzzy ART dynamics.

Category choice: For each input $\mathbf{I}$ and F_2 node j, the *choice function* T_j is defined by

$$T_j(\mathbf{I}) = \frac{|\mathbf{I} \wedge \mathbf{w}_j|}{\alpha + |\mathbf{w}_j|}, \tag{2}$$

where the fuzzy intersection $\wedge$ (Zadeh, 1965) is defined by

$$(\mathbf{p} \wedge \mathbf{q})_i \equiv \min(p_i, q_i) \tag{3}$$

and where the norm $|\cdot|$ is defined by

$$|\mathbf{p}| \equiv \sum_{i=1}^{M} |p_i|. \tag{4}$$

The system makes a *category choice* when at most one F_2 node can become active at a given time. The index J denotes the chosen category, where

$$T_J = \max\{T_j : j = 1 \ldots N\}. \tag{5}$$

When the Jth category is chosen, $y_J = 1$; and $y_j = 0$ for $j \neq J$.

Resonance or reset: *Resonance* occurs if the *match function* $|\mathbf{I} \wedge \mathbf{w}_J|/|\mathbf{I}|$ of the chosen category meets the vigilance criterion:

$$\frac{|\mathbf{I} \wedge \mathbf{w}_J|}{|\mathbf{I}|} \geq \rho. \tag{6}$$

Learning then ensues, as defined below. Otherwise *Mismatch reset* occurs, where the value of the choice function T_J is set to 0 for the duration of the input presentation. The search process continues until a chosen category J satisfies the matching criterion (6).

Learning: Once search ends, the weight vector $\mathbf{w}_J$ learns according to the equation

$$\mathbf{w}_J^{(\text{new})} = \beta(\mathbf{I} \wedge \mathbf{w}_J^{(\text{old})}) + (1-\beta)\mathbf{w}_J^{(\text{old})}. \tag{7}$$

Fast learning corresponds to setting $\beta = 1$. Using the fast learning and slow recoding option, we set $\beta = 1$ when J is an uncommitted node and take $\beta < 1$ after the category is committed.

Normalization by complement coding: Normalization of fuzzy ART inputs prevents category proliferation. The complement coded $F_0 \rightarrow F_1$ input $\mathbf{I}$ is the 2M-dimensional vector

$$\mathbf{I} = (\mathbf{a}, \mathbf{a}^c) \equiv (a_1, \ldots, a_M, a_1^c, \ldots, a_M^c), \tag{8}$$

where

$$a_i^c \equiv 1 - a_i. \tag{9}$$

A complement coded input is automatically normalized, because

$$|\mathbf{I}| = |(\mathbf{a}, \mathbf{a}^c)| = \sum_{i=1}^{M} a_i + (M - \sum_{i=1}^{M} a_i) = M. \tag{10}$$

With complement coding, the initial condition

$$w_{j1}(0) = \ldots = w_{j,2M}(0) = 1, \tag{11}$$

replaces the fuzzy ART initial condition (1).

2.2 ARTMAP Prediction and Search

Fuzzy ARTMAP incorporates two fuzzy ART modules ART_a and ART_b that are linked together via an inter-ART map field F^{ab}. The map field forms predictive associations between categories and realizes the ARTMAP *match tracking rule.*

ART_a and ART_b: Inputs to ART_a and ART_b are complement coded. For ART_a, $\mathbf{I} = \mathbf{A} = (\mathbf{a}, \mathbf{a}^c)$; and for ART_b, $\mathbf{I} = \mathbf{B} = (\mathbf{b}, \mathbf{b}^c)$ (Figure 3). For ART_a, $\mathbf{x}^a$ denotes the F_1^a output vector; $\mathbf{y}^a$ denotes the F_2^a output vector; and $\mathbf{w}_j^a$ denotes the jth ART_a weight vector. For ART_b, $\mathbf{x}^b$ denotes the F_1^b output vector; $\mathbf{y}^b$ denotes the F_2^b output vector; and $\mathbf{w}_k^b$ denotes the kth ART_b weight vector. For the map field, $\mathbf{x}^{ab}$ denotes the F^{ab} output vector, and $\mathbf{w}_j^{ab}$ denotes the weight vector from the jth F_2^a node to F^{ab}.

Map field activation: The map field F^{ab} receives input from either or both of the ART_a or ART_b category fields. A chosen F_2^a node J sends input to the map field F^{ab} via the weights $\mathbf{w}_J^{ab}$. An active F_2^b node K sends input to F^{ab} via one–to-one pathways between F_2^b and F^{ab}. If both ART_a and ART_b are active, then F^{ab} remains active only if ART_a predicts the same category as ART_b. The F^{ab} output vector $\mathbf{x}^{ab}$ obeys:

$$\mathbf{x}^{ab} = \begin{cases} \mathbf{y}^b \wedge \mathbf{w}_J^{ab} & \text{if the J}^{th}\ F_2^a \text{ node is active and } F_2^b \text{ is active} \\ \mathbf{w}_J^{ab} & \text{if the J}^{th}\ F_2^a \text{ node is active and } F_2^b \text{ is inactive} \\ \mathbf{y}^b & \text{if } F_2^a \text{ is inactive and } F_2^b \text{ is active} \\ \mathbf{0} & \text{if } F_2^a \text{ is inactive and } F_2^b \text{ is inactive.} \end{cases} \tag{12}$$

By (12), $\mathbf{x}^{ab} = \mathbf{0}$ if $\mathbf{y}^b$ fails to confirm the map field prediction made by $\mathbf{w}_J^{ab}$. Such a mismatch event triggers an ART_a search for a better category, as follows.

Match tracking: At the start of each input presentation ART_a vigilance ρ_a equals a baseline vigilance parameter $\overline{\rho_a}$. When a predictive error occurs, match tracking raises ART_a vigilance just enough to trigger a search for a new F_2^a coding node. ARTMAP detects a predictive error when

$$|\mathbf{x}^{ab}| < \rho_{ab}|\mathbf{y}^b|, \tag{13}$$

where ρ_{ab} is the map field vigilance parameter. A signal from the map field to the ART_a orienting subsystem causes ρ_a to "track the F_1^a match."

That is, ρ_a increases until it is slightly higher than the F_1^a match value $|\mathbf{A} \wedge \mathbf{w}_J^a||\mathbf{A}|^{-1}$. Then, since ART$_a$ fails to meet the matching criterion, the search for another F_2^a node begins.

Map field learning: Weights w_{jk}^{ab} in $F_2^a \rightarrow F^{ab}$ paths initially satisfy

$$w_{jk}^{ab}(0) = 1. \tag{14}$$

During resonance with the ART$_a$ category J active, $\mathbf{w}_J^{ab}$ approaches the map field vector $\mathbf{x}^{ab}$ as in (7). With fast learning, once J learns to predict the ART$_b$ category K, that association is permanent; i.e., $w_{JK}^{ab} = 1$ for all time.

3 Cascade ARTMAP

Prior knowledge of a problem domain could help a neural network in learning to solve the problem. Specifically, pre-existing symbolic rules can be used to initialize a neural network architecture before learning. Not only can rule insertion improve network learning efficiency, it also serves to provide knowledge that is not captured by training cases or that cannot be easily learned by a neural network, and thus improves the system predictive performance. In addition, incomplete or partially correct symbolic knowledge can be refined or enhanced through neural network learning algorithms. Rule insertion and refinement in neural networks therefore automates symbolic knowledge enhancement and repair.

A popular approach to rule insertion and refinement uses rules to initialize the architecture of a multi-layer neural network and refines the network using a backpropagation algorithm [8], [9], [20]. One major problem of using backpropagation networks (BP) to refine rule-based knowledge is the preservation of symbolic knowledge. Under the weight tuning process of a backpropagation algorithm, symbolic rules quickly lose their original meanings. In fact, large shifts in the meanings of hidden units can occur as a result of training [19].

Another major limitation of the BP approach is that the initial rule base has to be roughly complete, or else the initial network architecture created may not be sufficiently rich for dealing with the problem domain. As the standard backpropagation algorithm is not able to create additional nodes or connections dynamically during learning, a network initialized by a small set of rules may even have a lower chance of eventually learning the task. This problem was noted and partially solved by Lacher *et. al.*, who used virtual rules to create potential connections for learning [9]. However, in general, it is difficult to decide beforehand the virtual rules or connections desired. Tresp, Hollatz, and Ahmad [21] employed a learning algorithm that allowed creation of basis functions during learning. However, as their model only represents rules associating input attributes to output predictions, the network is not general enough to deal with rule-based domain theories involving intermediate attributes and rule chaining.

3.1 Rule Cascade Representation

Note that ARTMAP also handles a class of IF-THEN rules that map a set of input attributes directly to a disjoint set of output attributes. Symbolic rule-based knowledge, on the other hand, often involves *rule cascades* and *intermediate attributes*. A set of rules is said to form a *rule cascade* when a consequent of a rule also serves as an antecedent of another rule. Such attributes that have dual roles are usually called *intermediate attributes* in contrast to *input attributes* that only serve as antecedents and *output attributes* that only serve as consequents. Through intermediate attributes, firing of a rule may lead to the firing of another rule at a later stage of an inferencing process. Intermediate attributes and rule cascades are useful for feature abstraction and task decomposition so that only a small set of simple rules is needed at each level of the cascade.

The proposed solution to representing rule cascades here is Cascade ARTMAP that uses ARTMAP architecture but generalizes one-step prediction to multi-step inferencing. Cascade ARTMAP unifies the ARTMAP input attribute field F_1^a and output attribute field F_1^b in the sense that both F_1^a and F_1^b contain the input, output, and intermediate attributes. Consider the two rules below that form a simple two-level rule cascade:

Rule 1: IF A and B THEN C
Rule 2: IF C and D THEN E

where A, B, and D are input attributes; C is an intermediate attribute; and

E is an output attribute. All attributes (A, B, C, D, and E) are represented in both F_1^a and F_1^b (Figure 4). For Rule 1, an F_2^a category node is used to encode A and B, and is associated to an F_2^b node that predicts C. Likewise for Rule 2, an F_2^a node is used to encode C and D, and is associated to an F_2^b node predicting E. By unifying the input field (F_1^a) and the output field (F_1^b) of ARTMAP, several desired features of symbolic processing are obtained. Besides that rule insertion can be readily performed in Cascade ARTMAP, rule chaining and inferencing can also be performed as in production systems.

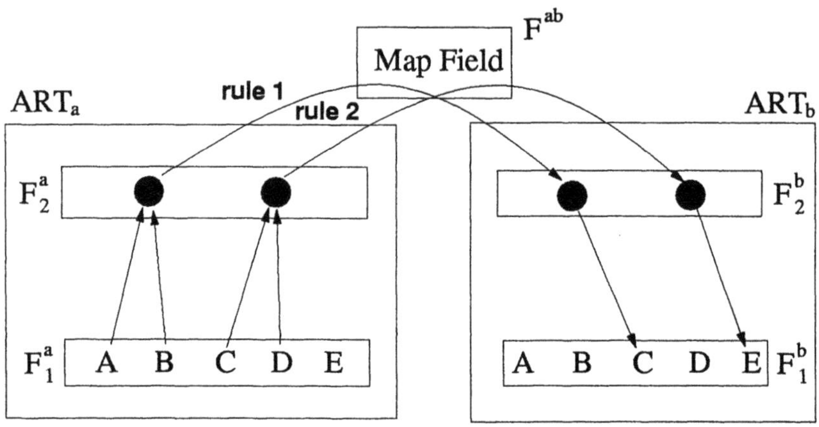

Figure 4. Cascade ARTMAP representation of the sample rule cascade.

3.2 Rule Insertion

As the knowledge structure of Cascade ARTMAP is compatible with rule-based knowledge representation, IF-THEN rules can be readily translated into the recognition categories of a Cascade ARTMAP system. Initializing a Cascade ARTMAP network with pre-existing rules before learning serves to set up the global solution structure. This helps to improve learning efficiency and predictive accuracy. Without rule insertion, Cascade ARTMAP performance reduces to that of fuzzy ARTMAP.

Rule insertion proceeds in two phases. The first phase parses all rules for attribute names to set up a *symbol table* in which each attribute in the rules has a unique entry. Based on the symbol table, the second phase translates each rule into two 2M-dimensional vectors **A** and **B**, where M is the total number of attributes in the symbol table, as inputs to the Cascade ARTMAP ART$_a$ and ART$_b$ modules. Given a rule of the following format,

$$\begin{array}{ll} \text{IF} & x_1, x_2, \ldots, x_m, \neg\bar{x}_1, \neg\bar{x}_2, \ldots, \neg\bar{x}_{\bar{m}} \\ \text{THEN} & y_1, y_2, \ldots, y_n, \neg\bar{y}_1, \neg\bar{y}_2, \ldots, \neg\bar{y}_{\bar{n}} \end{array}$$

where $x_1, \ldots, x_m$ and $y_1, \ldots, y_n$ are positive attributes, and $\bar{x}_1, \ldots, \bar{x}_{\bar{m}}$ and $\bar{y}_1, \ldots, \bar{y}_{\bar{n}}$ preceded by the logical NOT operator $\neg$ are negative attributes, the algorithm derives the pair of vectors

$$\mathbf{A} = (\mathbf{a}, \mathbf{a}^c) \quad \text{and} \quad \mathbf{B} = (\mathbf{b}, \mathbf{b}^c) \tag{15}$$

such that for each index $j = 1, \ldots, M$,

$$(a_j, a_j^c) = \begin{cases} (1,0) & \text{if } e_j = x_i \text{ for some } i \in \{1, \ldots, m\} \\ (0,1) & \text{if } e_j = \bar{x}_i \text{ for some } i \in \{1, \ldots, \bar{m}\} \\ (0,0) & \text{otherwise} \end{cases} \tag{16}$$

and

$$(b_j, b_j^c) = \begin{cases} (1,0) & \text{if } e_j = y_i \text{ for some } i \in \{1, \ldots, n\} \\ (0,1) & \text{if } e_j = \bar{y}_i \text{ for some } i \in \{1, \ldots, \bar{n}\} \\ (0,0) & \text{otherwise} \end{cases} \tag{17}$$

where e_j is the jth attribute in the symbol table. Note that complement coding is employed above for encoding both the positive and negative attributes. If the rules contain no negative attribute, the complement vectors $\mathbf{a}^c$ and $\mathbf{b}^c$ may be eliminated.

The vector pairs derived from the rules are then used as training patterns to initialize a Cascade ARTMAP network. The network learning and inferencing algorithms will be described in subsequent sections. It suffices to note at this point that each distinct vector $\mathbf{A}$ is translated into a recognition category in ART$_a$ and likewise each distinct vector $\mathbf{B}$ is translated into a recognition category in ART$_b$. Given a pair of vectors $\mathbf{A}$ and $\mathbf{B}$ derived from a rule, their respective recognition categories are associated through the map field. During network initialization, the vigilance parameters ρ_a and ρ_b are each set to 1 to ensure that only identical attribute vectors are grouped into one recognition category. Contradictory symbolic rules are detected during rule insertion when identical input attribute vectors are associated with distinct output attribute vectors. The detection is achieved through a *perfect mismatch* phenomenon, in which the system tries to raise ART_a vigilance ρ_a above 1 in response to a mismatch in ART_b.

3.3 Rule Chaining and Inferencing

A symbolic production rule system consists of three main components: a *working memory* component, a *rule* or *production* component, and an external inference engine or *interpreter*. The interpreter repeatedly performs a three-phase cycle, consisting of a *match* phase, a *select* phase, and an *execute* phase. In the match phase, the interpreter compares the antecedent set of each rule against the working memory content. Rules with completely matched antecedents are included into a *conflict* set. In the select phase, a single rule is selected from the conflict set using some strategies. If the conflict set is empty, the system halts. Otherwise, in the execute phase, the interpreter adds the consequent(s) of the selected rule to the working memory.

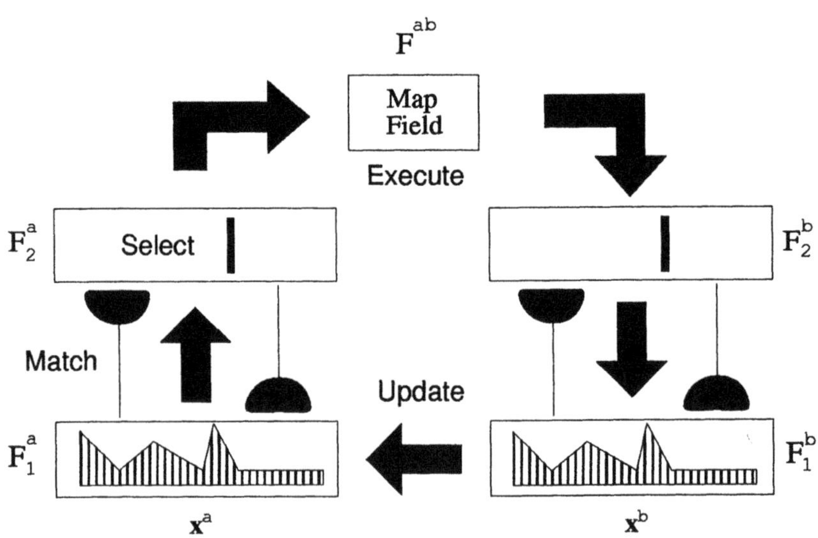

Figure 5. Cascade ARTMAP rule chaining and inferencing: $\mathbf{x}^a$ represents the system's memory state and accumulates attribute values during multi-step inferencing.

In Cascade ARTMAP, the attribute fields F_1^a and F_1^b can be identified as the working memory component (Figure 5). F_1^a maintains the current memory state $\mathbf{x}^a$ and provides the antecedents for condition matching and rule firing. F_1^b stores the next memory state $\mathbf{x}^b$ derived through a rule firing. The rule component is implemented by the two category fields F_2^a and F_2^b, the map field F^{ab}, and their interconnections. The match, select, and execute three-phase cycle is performed without an external interpreter. In the match phase, a choice function T_j^a is computed for each F_2^a category node (rule) based on the memory state $\mathbf{x}^a$. With parallel implementation, the match phase can be performed in a single activation process. The select phase is realized by a winner-take-all interaction among all F_2^a nodes in which the F_2^a node with the largest choice function T_j^a is identified. If the selected node (rule) does not satisfy the ART$_a$ vigilance constraint, the system goes through another round of memory search to select another F_2^a node that satisfies the ART$_a$ vigilance criterion. If no such node exists, the system halts. Otherwise, in the execute phase, the consequent(s) of the selected rule is(are) read out into F_1^b. Note that exact match is not required for a rule to be fired as long as it satisfies the ART$_a$ vigilance criterion. At the end of the cycle, the new memory state $\mathbf{x}^b$ is used to update $\mathbf{x}^a$ in F_1^a to prepare for the next inferencing cycle. For the sample rule cascade (Figure 4), the input attribute set {A,B,D} activates F_2^a node J_1 that infers C. Through the memory update process, C is fed back from F_1^b to F_1^a. The memory state $\mathbf{x}^b$ which contains {A,B,C,D} then activates J_2 in the next inferencing cycle that derives E.

3.4 Learning and Rule Refinement

Learning in Cascade ARTMAP is more complicated than that in fuzzy ARTMAP as a chain of rule firing is involved in making a prediction. The proposed solution is a backtracking algorithm that identifies all rules (F_2^a nodes) responsible for making a prediction by tracing from the last rule fired. Specifically, if J is the last F_2^a node selected which makes the prediction, the algorithm identifies a *precursor* set $\Psi(J)$ that contains node J and all F_2^a nodes that result in the firing of node J. The backtracking occurs in the direction of $F_2^a \rightarrow F_1^a \rightarrow F_1^b \rightarrow F_2^b \rightarrow F^{ab} \rightarrow F_2^a$. For example in Figure 4, the backtracking algorithm traces from J_2 in F_2^a to its antecedents {C,D} in F_1^a. It then checks that C in F_1^b is an intermediate attribute activated by K_1 in F_2^b, and finally traces to J_1 in F_2^a. The backtracking stops at J_1 as its antecedents are all input attributes. The precursor set of the F_2^a node J_2, $\Psi(J_2)$ is thus evaluated to be $\{J_1, J_2\}$.

If the prediction made by node J is correct, for each F_2^a node (rule) j in the precursor set $\Psi(J)$, the weight vector (antecedent set) $\mathbf{w}_j^a$ is reduced towards its fuzzy intersection with the F_1^a activity vector $\mathbf{x}^a$. In the binary pattern and fast learning case, a fired rule learns to ignore those features that are absent in the current input. This results in a generalization by reducing the number of features the rule attends to.

A more complicated situation occurs when a prediction error is encountered. With a long chain of rule firing, blame assignment can be difficult as it is unclear which rule in the inferencing path causes the error. To handle prediction mismatch, a *mini-match tracking* process raises the ART$_a$ vigilance ρ_a by slightly more than the *minimum* match achieved by the fired rules. Mini-match tracking is equivalent to the parallel match tracking mechanism used in Fusion ARTMAP [1]. This method inhibits the F_2^a node with the minimum match from firing again for the current input. The assumption is that the rule with the worst match is most likely to be the one which causes the prediction error. The system then goes through another round of memory search and inferencing with a higher vigilance until a resonance is achieved.

3.5 Cascade ARTMAP Algorithm

As an on-line real-time system, Cascade ARTMAP needs not separate learning and performance phases, i.e., the system functions in response to the current input environment. Given an F_1^a input vector, Cascade ARTMAP undergoes a series of prediction loops until either an uncommitted F_2^a node is selected (which means no prediction), or a correct prediction is made by a committed F_2^a node (as in fuzzy ARTMAP). In each prediction loop, Cascade ARTMAP accumulates intermediate attribute values through a series of inferencing cycles until one or more output attributes are derived. The Cascade ARTMAP dynamics, as illustrated in Figures 6 and 7, are formalized below. An *A then B* paradigm is used in which the ART$_a$ input vector $\mathbf{A}$ is processed before the ART$_b$ input vector $\mathbf{B}$.

Activity vectors: Let $\mathbf{A}$ and $\mathbf{B}$ denote the F_1^a and F_1^b input vectors respectively. Let

$$\mathbf{x}^a \equiv (\mathbf{x}_i^a, \mathbf{x}_h^a, \mathbf{x}_o^a, \mathbf{x}_i^{ac}, \mathbf{x}_h^{ac}, \mathbf{x}_o^{ac}) \tag{18}$$

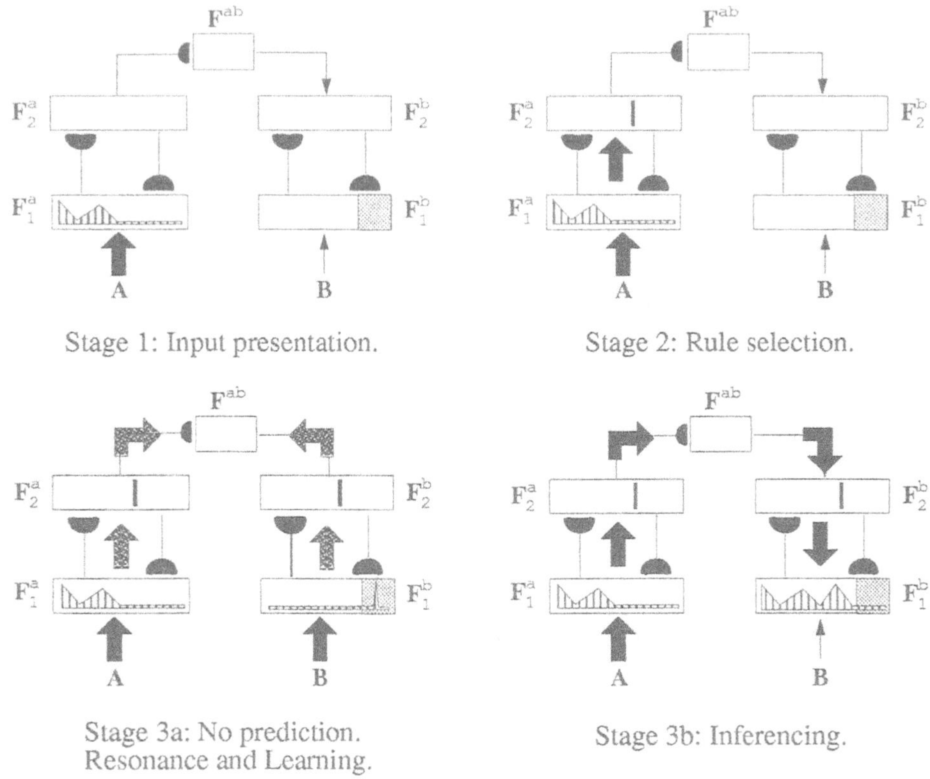

Figure 6. Cascade ARTMAP algorithm stages 1, 2, and 3. The shaded sub-fields of F_1^b represent output attributes.

and

$$\mathbf{x}^b \equiv (\mathbf{x}_i^b, \mathbf{x}_h^b, \mathbf{x}_o^b, \mathbf{x}_i^{bc}, \mathbf{x}_h^{bc}, \mathbf{x}_o^{bc}) \tag{19}$$

denote the 2M-dimensional F_1^a and F_1^b activity vectors respectively, where $\mathbf{x}_i^a$ and $\mathbf{x}_i^b$ denote the M_i-dimensional input attribute vectors; $\mathbf{x}_h^a$ and $\mathbf{x}_h^b$ denote the M_h-dimensional intermediate attribute vectors; $\mathbf{x}_o^a$ and $\mathbf{x}_o^b$ denote the M_o-dimensional output attribute vectors; and $\mathbf{x}_i^{ac}$, $\mathbf{x}_i^{bc}$, $\mathbf{x}_h^{ac}$, $\mathbf{x}_h^{bc}$, $\mathbf{x}_o^{ac}$, and $\mathbf{x}_o^{bc}$ denote the respective complement attribute vectors. $\mathbf{x}^a$ and $\mathbf{x}^b$ are also known as memory state vectors. Let $\mathbf{y}^a$ and $\mathbf{y}^b$ denote the F_2^a and F_2^b activity vectors respectively. Let $\mathbf{x}^{ab}$ denote the map field F^{ab} activity vector.

Weight vectors: Let $\mathbf{w}_j^a$ and $\mathbf{w}_j^b$ denote the 2M-dimensional weight vectors of the jth category node in F_2^a and F_2^b respectively. Let $\mathbf{w}_j^{ab}$ denote the weight vector from the jth F_2^a node to F^{ab}. Initially, the weight vec-

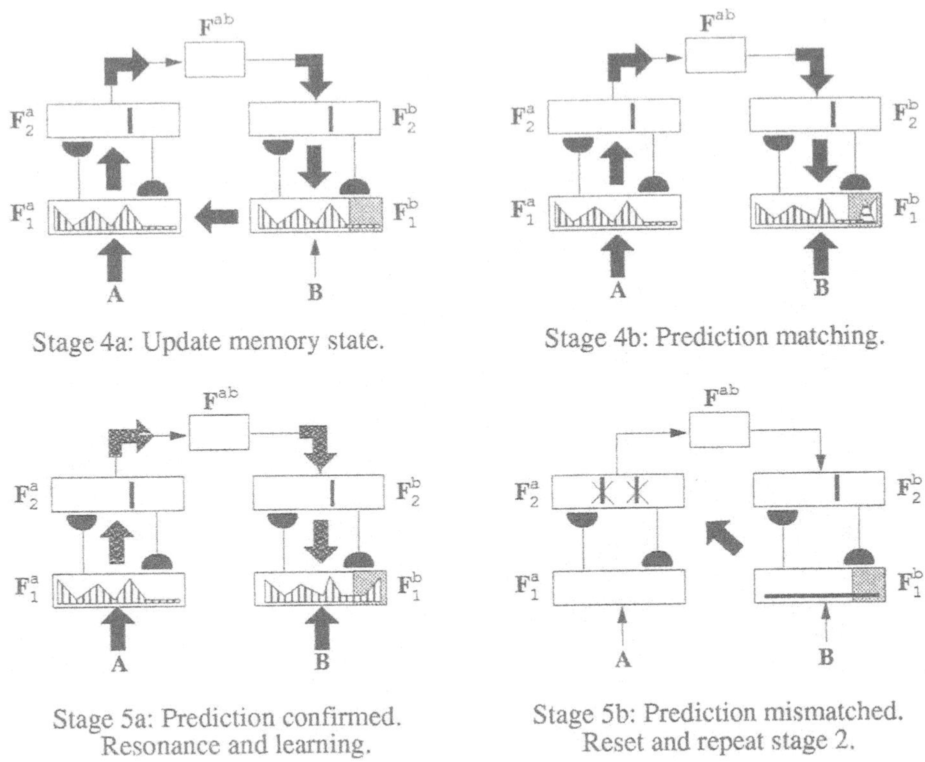

Figure 7. Cascade ARTMAP algorithm stages 4 and 5. The shaded sub-fields of F_1^b represent output attributes.

tors contain all 1's. This implies that all category nodes are uncommitted and all F_2^a nodes are not associated with any prediction.

Scope vectors: Let $\mathbf{S}_j$ denote the 2M-dimensional scope vector of the jth category node in F_2^a. A scope vector identifies the attributes relevant to an F_2^a node and allows a more accurate computation of its match function. For an uncommitted F_2^a node j, the scope vector $\mathbf{S}_j \equiv (\mathbf{s}_j, \mathbf{s}_j)$ is defined by

$$s_{ji} = \begin{cases} 1 & \text{if } i \text{ indexes an input attribute} \\ 0 & \text{otherwise.} \end{cases} \tag{20}$$

For an F_2^a node j created by an inserted rule, the scope vector $\mathbf{S}_j \equiv (\mathbf{s}_j, \mathbf{s}_j)$ is defined by

$$s_{ji} = \begin{cases} 1 & \text{if } i \text{ indexes an attribute at a level previous to} \\ & \text{the rule's consequent(s) in the rule cascade} \\ 0 & \text{otherwise.} \end{cases} \tag{21}$$

Parameters: Cascade ARTMAP dynamics are determined by the choice parameters $\alpha_a > 0$ and $\alpha_b > 0$; the learning rates $\beta_a \in [0,1]$ and $\beta_b \in [0,1]$; and the vigilance parameters $\rho_a \in [0,1]$ and $\rho_b \in [0,1]$. During network initialization, the network learns the patterns derived from rules using $\rho_a = \rho_b = 1$ such that each distinct attribute vector creates a category. During network refinement, the system learns example patterns using $\rho_b = 1$ for output classification and $\rho_a < 1$ to allow input generalization.

Stage 1 (Input presentation): At the beginning of an input presentation, the ART$_a$ vigilance ρ_a equals a baseline vigilance value $\bar{\rho_a}$. F_1^a contains the input vector **A**:

$$\mathbf{x}^a = \mathbf{A}. \tag{22}$$

Stage 2 (Rule selection): Given the memory state vector $\mathbf{x}^a$, for each F_2^a node j, the choice function T_j^a is defined by

$$T_j^a = \frac{|\mathbf{x}^a \wedge \mathbf{w}_j^a|}{\alpha_a + |\mathbf{w}_j^a|}, \tag{23}$$

where the fuzzy AND operator $\wedge$ is defined by

$$(\mathbf{p} \wedge \mathbf{q})_i \equiv min(p_i, q_i), \tag{24}$$

and the norm $|.|$ is defined by

$$|\mathbf{p}| = \sum_i |p_i| \tag{25}$$

for vectors **p** and **q**. The system is said to make a *category choice* when at most one F_2^a node can become active at a given time. The category choice is indexed at J, where

$$T_J^a = \max\{T_j^a : \text{for all } F_2^a \text{ node } j\}. \tag{26}$$

If more than one T_j^a is maximal, the F_2^a category node j with the smallest index is chosen. In particular, nodes become committed in order $j =$

$1, 2, 3, \ldots$. When the Jth category is chosen, $y_J^a = 1$; and $y_j^a = 0$ for $j \neq J$.

Resonance occurs if the match function m_J^a of the selected node J meets the vigilance criterion:

$$m_J^a = \frac{|\mathbf{x}^a \wedge \mathbf{w}_J^a \wedge \mathbf{S}_J|}{|\mathbf{x}^a \wedge \mathbf{S}_J|} \geq \rho_a, \tag{27}$$

where the generalized fuzzy AND operation $\wedge$ is defined by

$$(\wedge_{j=1}^N \mathbf{p}_j)_i \equiv \min(p_{1i}, \ldots, p_{Ni}), \tag{28}$$

for vectors $\mathbf{p}_1, \ldots, \mathbf{p}_N$, and the norm $|.|$ is defined as in (25). Otherwise mismatch reset occurs in which the value of the choice function T_J^a is set to 0 for the duration of the input presentation to prevent persistent selection of the same category during search. Stage 2 then repeats to select another new index J.

Stage 3a (No prediction): If the selected F_2^a node J has no prediction, i.e.,

$$w_{Jk}^{ab} = 1 \quad \text{for all } F^{ab} \text{ node } k, \tag{29}$$

each F_2^a node j in the precursor set $\Psi(J)$ learns the F_1^a activity pattern according to the equation

$$\mathbf{w}_j^{a(\text{new})} = (1 - \beta_a)\mathbf{w}_j^{a(\text{old})} + \beta_a(\mathbf{x}^a \wedge \mathbf{w}_j^{a(\text{old})}). \tag{30}$$

If the input vector $\mathbf{B}$ is present, a category node is selected in F_2^b as in stage 2. The selected F_2^b node K learns the F_1^b pattern according to the equation

$$\mathbf{w}_K^{b(\text{new})} = (1 - \beta_b)\mathbf{w}_K^{b(\text{old})} + \beta_b(\mathbf{x}^b \wedge \mathbf{w}_K^{b(\text{old})}). \tag{31}$$

The F_2^a category node J is then associated to the F_2^b node K through the inter-ART map field:

$$w_{Jk}^{ab} = \begin{cases} 1 & \text{if } k = K \\ 0 & \text{otherwise.} \end{cases} \tag{32}$$

After which, the system halts.

Stage 3b (Inferencing): If the selected F_2^a node J has learned to make a prediction, i.e., (29) does not hold, its weight vector $\mathbf{w}_J^{ab}$ activates F^{ab}.

The F^{ab} activity vector $\mathbf{x}^{ab}$ is defined by

$$\mathbf{x}^{ab} = \mathbf{w}_J^{ab}. \tag{33}$$

Once the map field is active, F_2^b is activated through the 1-to-1 pathway between F^{ab} and F_2^b . For each F_2^b node k, the choice function T_k^b is defined by

$$T_k^b = x_k^{ab}. \tag{34}$$

The system again makes a category choice indexed at K where

$$T_K^b = \max\{T_k^b\text{: for all } F_2^b \text{ node } k\}. \tag{35}$$

When the Kth category is chosen, $y_K^b = 1$; and $y_k^b = 0$ for $k \neq K$. The activated F_2^b node K then performs a top-down priming on F_1^b:

$$\mathbf{x}^b = \mathbf{w}_K^b. \tag{36}$$

When F_1^b is activated by a category choice in F_2^b, the termination condition is checked by computing a goal signal g:

$$g = \sum_{i=1}^{M_o}(x_{oi}^b + x_{oi}^{bc}). \tag{37}$$

A conclusion is reached whenever any output attribute is made known, i.e., $g > 0$.

Stage 4a (Update memory state): If a conclusion is not reached, i.e., $g = 0$, the memory state vector $\mathbf{x}^a$ is updated with $\mathbf{x}^b$ by the equation

$$\mathbf{x}^{a(\text{new})} = \mathbf{x}^{a(\text{old})} \vee \mathbf{x}^{b(\text{old})}, \tag{38}$$

where the fuzzy OR operation $\vee$ is defined by

$$(\mathbf{p} \vee \mathbf{q})_i \equiv max(p_i, q_i) \tag{39}$$

for vectors $\mathbf{p}$ and $\mathbf{q}$. The inferencing cycle then repeats from stage 2.

Stage 4b (Prediction matching): If a conclusion is reached, i.e., $g > 0$, the match function m_K^b of the prediction $\mathbf{x}^b$ and the F_1^b input vector $\mathbf{B}$ is computed by

$$m_K^b = \frac{|\mathbf{B} \wedge \mathbf{x}^b|}{|\mathbf{B}|}. \tag{40}$$

Stage 5a (Resonance): If the prediction match satisfies the ART_b vigilance criterion ($m_K^b \geq \rho_b$), resonance occurs. The activated F_2^a and F_2^b nodes learn the template patterns in their respective modules as in (30) and (31) respectively. After learning, the system halts.

Stage 5b (Match tracking): A prediction mismatch triggers a match tracking process. Using mini-match tracking, a node j is identified which has the minimum match function value among all nodes in $\Psi(J)$. The choice function T_j^a of the node j is set to zero during the input presentation. The ART_a vigilance ρ_a is raised to slightly greater than the match achieved by the node j_m:

$$\rho_a^{(\text{new})} = \max\{\rho_a^{(\text{old})}, \min\{m_j^a | j \in \Psi(J)\} + \epsilon\}. \tag{41}$$

Perfect mismatch occurs when the system attempts to increase ρ_a above 1. A perfect match in ART_a ($\rho_a = 1$) with a ART_b mismatch indicates the existence of contradictory knowledge where identical antecedent sets are associated with different consequents. After match tracking, a new prediction loop then repeats from stage 2.

3.6 Rule Extraction

Rules can be derived more readily from an ARTMAP network than from a backpropagation network, in which the roles of hidden units are usually not explicit. In an ARTMAP network, each node in the F_2^a field represents a recognition category of ART_a input patterns. Through the inter-ART map field, each such node is associated to an ART_b category in the F_2^b field, which in turn encodes a prediction. Learned weight vectors, one for each F_2^a node, constitute a set of rules that link antecedents to consequences (Figure 1). The number of rules equals the number of F_2^a nodes that become active during learning.

As large databases typically cause ARTMAP to generate too many rules to be of practical use. The goal of the rule extraction task is to select a small set of highly predictive category nodes and to describe them in a comprehensible form. To evaluate a category node, a *confidence factor* that measures both *usage* and *accuracy* is computed. Removal of low confidence recognition categories created by atypical examples produces smaller networks. Removal of redundant weights in a category node's weight vector reduces the number of antecedents in the corresponding rule.

3.6.1 Rule Pruning

The rule pruning algorithm derives a confidence factor for each F_2^a category node in terms of its usage frequency in a *training* set and its predictive accuracy on a *predicting* set. As Cascade ARTMAP generalizes ARTMAP one-step prediction process to multi-step inferencing, an input pattern makes use of a set of F_2^a category nodes in Cascade ARTMAP in contrast to a single F_2^a node in fuzzy ARTMAP. For evaluating *usage* and *accuracy*, each F_2^a category node j maintains three counters: an encoding counter c_j, that records the number of training set patterns encoded by node j; a predicting counter p_j, that records the number of predicting set patterns predicted by node j; and a success counter s_j, that records the number of predicting set patterns predicted correctly by node j.

For each training set pattern, the encoding counter (c_j) of each F_2^a node j in the precursor set $\Psi(J)$, where J is the last F_2^a node (rule) fired that makes the prediction, is increased by 1. For each predicting set pattern, the predicting counter (p_j) of each F_2^a node j in the precursor set $\Psi(J)$ is increased by 1. If the prediction is correct, the success counter (s_j) of each F_2^a node j in the precursor set $\Psi(J)$ is increased by 1. Based on the encoding, predicting, and success counter values, the *usage* (U_j) and the accuracy (A_j) of an F_2^a node j are computed by

$$U_j = c_j / \max\{c_k\text{: for all } F_2^a \text{ node } k\} \tag{42}$$

and

$$A_j = P_j / \max\{P_k\text{: for all } F_2^a \text{ node } k\}, \tag{43}$$

where P_j, the percent of the predicting set pattern predicted correctly by node j, is computed by

$$P_j = s_j / p_j. \tag{44}$$

U_j and A_j are then used to compute the confidence factor of node j by the equation

$$CF_j = \gamma U_j + (1 - \gamma) A_j, \tag{45}$$

where $\gamma \in [0, 1]$ is a weighting factor. After confidence factors are determined, recognition categories can be pruned from the network using one of following strategies.

Threshold Pruning - This is the simplest type of pruning where the F_2^a nodes with confidence factors below a given threshold τ are removed from the network. A typical setting for τ is 0.5. This method is fast and provides a first cut elimination of unwanted nodes. To avoid over-pruning, it is sometimes useful to specify a minimum number of recognition categories to be preserved in the system.

Local Pruning - Local pruning removes recognition categories one at a time from an ARTMAP network. The baseline system performance on the training and the predicting sets is first determined. Then the algorithm deletes the recognition category with the lowest confidence factor. The category is replaced, however, if its removal degrades system performance on the training and predicting sets.

A variant of the local pruning strategy updates baseline performance each time a category is removed. This option, called *hill-climbing*, gives slightly larger rule sets but better predictive accuracy. A hybrid strategy first prunes ARTMAP using threshold pruning and then applies local pruning on the remaining smaller set of rules.

3.6.2 Antecedent Pruning

During rule extraction, a non-zero weight to an F_2^a category node translates into an antecedent in the corresponding rule. The antecedent pruning procedure calculates an error factor for each antecedent in each rule based on its performance on the training and predicting sets. When a rule (F_2^a node) J makes a prediction error, for each F_2^a node j in the precursor set $\Psi(J)$, each antecedent of the rule j that also appears in the current memory state has its error factor increased in proportion to the smaller of its magnitudes in the rule and in the memory state vector $\mathbf{x}^a$. After the error factor for each antecedent is determined, a local pruning strategy, similar to the one for rules, removes redundant antecedents.

4 DNA Promoter Experiments

Promoters are short nucleotide sequences that occur before genes and serve as binding sites for the enzyme RNA polymerase during gene transcription. Identifying promoters is thus an important step in locating genes in DNA sequences. One major approach to DNA matching or sequence comparison concerns with the alignment of DNA sequences. Sequence alignment is usually performed by computing a match function which rewards matches and penalizes mismatches, insertions, and deletions [22]. This can be done by dynamic programming which can be computationally expensive for multiple sequences. Consensus sequence analysis solves the problem of aligning multiple sequences by identifying functionally important sequence features that are conserved in the DNA sequences. For example, consensus patterns of promoter sequences can be identified at the protein binding sites. Besides statistical methods reported in the biological literature, machine learning and information theoretic techniques are also being used for DNA matching and recognition [7], [10].

The promoter data set [11] used in the Cascade ARTMAP experiments consists of 106 patterns, half of which are positive instances (promoters). Although larger sets of promoter data are available, this version of the promoter data set is used here to allow a direct comparison with the results of the others. Each DNA pattern represents a 57-position window, with the leftmost 50 window positions labeled -50 to -1 and the rightmost seven labeled 1 to 7 (Figure 8). Each position is a nominal feature which takes one of the four nucleotide values {A, G, T, C}. There is no missing feature value. Using local representation, each 57-position pattern is expanded into a 228-bit nucleotide-position string.

The promoter data set and an imperfect domain theory have been used to evaluate a hybrid learning system called Knowledge Based Artificial Neural Network (KBANN) [20]. The imperfect domain theory (Table 1), if requires exact match, only classifies half of the 106 cases correctly. The KBANN theory refinement procedure translates the imperfect theory into a feedforward network, adds links to make the network fully connected between layers, and trains the network using a backpropagation algorithm. Simulation results showed that by incorporating the domain the-

Table 1. A rule-based theory for classifying promoters. It consists of 14 rules and a total of 83 antecedents. The antecedent notation T@-36 indicates the nucleotide value T in position -36.

promoter	:-	conformation, contact.
contact	:-	minus_35, minus_10.
minus_35	:-	C@-37, T@-36, T@-35, G@-34, A@-33, C@-32.
minus_35	:-	T@-36, T@-35, G@-34, C@-32, A@-31.
minus_35	:-	T@-36, T@-35, G@-34, A@-33, C@-32, A@-31.
minus_35	:-	T@-36, T@-35, G@-34, A@-33, C@-32.
minus_10	:-	T@-14, A@-13, T@-12, A@-11, A@-10, T@-9.
minus_10	:-	T@-13, A@-12, A@-10, T@-8.
minus_10	:-	T@-13, A@-12, T@-11, A@-10, A@-9, T@-8.
minus_10	:-	T@-12, A@-11, T@-7.
conformation	:-	C@-47, A@-46, A@-45, T@-43, T@-42, A@-40, C@-39, G@-22, T@-18, C@-16, G@-8, C@-7, G@-6, C@-5, C@-4, C@-2, C@-1.
conformation	:-	A@-45, A@-44, A@-41.
conformation	:-	A@-49, T@-44, T@-27, A@-22, T@-18, T@-16, G@-15, A@-1.
conformation	:-	A@-45, A@-41, T@-28, T@-27, T@-23, A@-21, A@-20, T@-17, T@-15, T@-4.

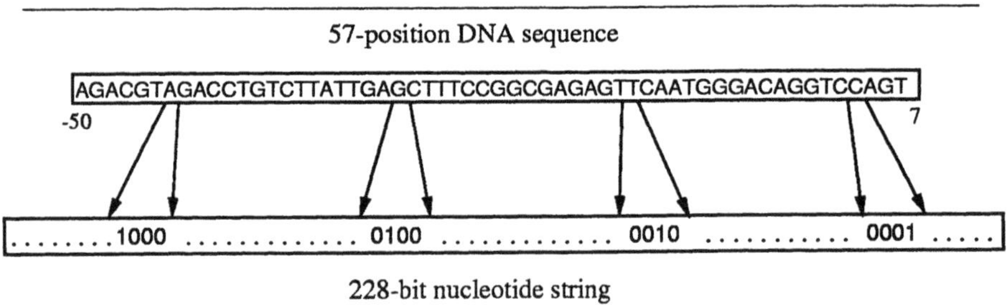

Figure 8. A 57-position DNA sequence. Each position takes one of the four nucleotide values {A,G,T,C}. Using local representation, each DNA sequence is expanded into a 228-bit nucleotide string. This version of 106-case promoter data set, obtained from the UCI machine learning database repository, contains no missing value.

Table 2. Performance of fuzzy ARTMAP, Cascade ARTMAP, and Cascade ARTMAP rules on the promoter data set comparing with the symbolic learning algorithm ID-3, the KNN system, consensus sequence analysis, the backpropagation network, the KBANN system, and the NOFM rules.

Systems	# Nodes/ Rules	# Antecedent	Error (%)
ID-3	-	-	17.9
KNN (K=3)	105	-	12.3
Consensus Sequences	-	-	11.3
Backpropagation Network	16	-	7.5
Fuzzy ARTMAP	20.6	-	6.5
KBANN	16	-	2.9
Cascade ARTMAP	13+15.9	-	2.0
NOFM rules	12	100	3.8
Cascade ARTMAP rules	19.5	53.1	3.0

ory, KBANN outperformed many learning/recognition systems, including consensus sequence analysis [13], K Nearest Neighbor (KNN), ID-3 symbolic learning algorithm [14], and backpropagation network trained purely from examples [20] (Table 2).

In Cascade ARTMAP experiments, the first two rules of the domain theory are combined into a single rule:

promoter :- conformation, minus_35, minus_10.

Besides providing a slight improvement in system predictive accuracy, the elimination of attribute *contact* reduces Cascade ARTMAP network complexity and produces simpler rule sets.

Cascade ARTMAP simulation is performed with parameter values $\alpha_a = \alpha_b = 2$ and $\beta_a = \beta_b = 1$, determined empirically. The input patterns are not complement coded as they already have a uniform norm of 57. In each simulation, Cascade ARTMAP is initialized with the domain theory, trained on 96 patterns selected randomly, and tested on the remaining 10 patterns. To use a voting strategy, Cascade ARTMAP is trained in several simulation runs using different orderings of the training set. For each test case, voting across 20 runs yields a final prediction. An averaging technique similar to voting was also used in the KBANN system [20].

Table 2 compares the performance of fuzzy ARTMAP and Cascade ARTMAP, averaged over 20 simulations, with other alternative systems. Among the systems that do not incorporate *a priori* symbolic knowledge, fuzzy ARTMAP (Cascade ARTMAP without rule insertion) achieves the lowest error rate. While the KBANN system and Cascade ARTMAP both obtain significant improvement in predictive performance by incorporating rules, Cascade ARTMAP produces a lower error rate than KBANN. In addition to the 13 inserted rules, an average of 15.9 recognition nodes (rules) are created.

In each simulation, rules are also extracted from the trained Cascade ARTMAP network. Due to the small data set size, confidence factors are computed solely based on *usage*. Threshold pruning with threshold $\tau = 0.01$ is applied, followed by the rule and antecedent pruning procedures using the local pruning strategy. Comparing predictive performance, rules extracted from Cascade ARTMAP are still slightly more accurate than the NOFM rules extracted from KBANN [18], [19]. While the Cascade ARTMAP rule sets contain more rules than the NOFM rule sets, the number of antecedents is almost half of that of the NOFM rule sets.

The promoter rules formulated by Cascade ARTMAP are similar in form to the consensus sequences derived by conventional statistical methods. However, whereas consensus sequences are used with an exact match condition, Cascade ARTMAP rules are based on competitive activation and do not require exact match in antecedents. Through the approximate matching property, the number of nucleotides used to identify a promoter is usually small (at most four in this case). By contrast, the consensus sequences, obtained by noting the positions with the same base in greater than 50% of the promoter patterns [12], used a minimum of twelve nucleotides.

Table 3 shows a sample set of refined promoter rules extracted from Cascade ARTMAP. *Conformation* has been dropped as a condition for promoters, so are the four rules defining it. All the minus_35 and minus_10 rules are preserved, but have been refined to refer to only two salient nucleotide bases. Two new rules for identifying promoters are created, which contain features of *minus_35* and *conformation*. These two rules

Table 3. A set of promoter rules extracted from Cascade ARTMAP. The set consists of 19 rules and a total of 46 antecedents. The real number associated with each rule represents the rule's confidence factor.

promoter (1.00)	:-	minus_35, minus_10.
promoter (0.31)	:-	A@-45, G@-34.
promoter (0.22)	:-	G@-34, T@-25, T@-18.
minus_35 (0.41)	:-	G@-34, C@-32.
minus_35 (0.34)	:-	T@-36, T@-35.
minus_35 (0.22)	:-	A@-33, C@-32.
minus_35 (0.03)	:-	T@-36, C@-32.
minus_10 (0.44)	:-	A@-12, T@-8.
minus_10 (0.31)	:-	A@-13, T@-9.
minus_10 (0.19)	:-	A@-11, T@-7.
minus_10 (0.06)	:-	A@-9, T@-8.
non-promoter (0.19)	:-	A@5.
non-promoter (0.16)	:-	A@-49, C@6, G@.7
non-promoter (0.16)	:-	A@7.
non-promoter (0.16)	:-	T@-23.
non-promoter (0.12)	:-	A@-15, T@1.
non-promoter (0.12)	:-	C@-46, G@-26.
non-promoter (0.06)	:-	T@-34, T@-33, C@-27, T@-26, G@5.
non-promoter (0.03)	:-	A@-45, T@-44, G@-42, T@-29, A@-24, T@-7, A@6, G@7.

are believed to compensate for the elimination of *conformation*. Eight non-promoter rules are created. They are slightly more irregular due to the randomness of non-promoters.

The confidence factor attached to each ARTMAP rule provides another dimension for interpreting the rule. By having a confidence factor of 1, the first promoter rule is very frequently used and thus important. It is activated by different combinations of minus_35 and minus_10 rules, each individually does not have a high usage. The two new promoter rules are roughly of equal importance but are not as heavily used as the first promoter rule. The first three minus_35 rules are more highly utilized than the last minus_35 rule. A similar pattern is observed for the minus_10 rules. The non-promoter rules have lower and less contrasting confidence values. The first four non-promoter rules nevertheless seem slightly more important. The last two non-promoter rules have the least confidences, and could be dropped with little degradation of overall performance.

In contrast, the promoter rules extracted by the NOFM algorithm consists of only 9 rules but contains 83 countable antecedents [19]. Moreover, the rules make use of several complex constructs, including NOFM, a counting function "nt", addition, subtraction, multiplication, and comparison of real numbers. Also, the NOFM rules involve seven nucleotide ambiguity codes, and have already employed a compressed format for representing adjacent nucleotide bases to simplify rules. Comparing complexity, ARTMAP rules are much cleaner and easier to interpret. More importantly, by preserving the symbolic rule form during learning, the extracted rules are identical in form and can be compared directly with the original rules. Furthermore, the use of confidence factors enables ranking of rules. This is particularly important to human experts in analyzing the rules.

5 Conclusion

This chapter has presented supervised Adaptive Resonance Theory systems in the perspective of symbolic knowledge processing. The inherent characteristics of the supervised ART systems, most notably, the fast and incremental learning capabilities and the compatibility with rule-based knowledge, give rise to a computing paradigm fundamentally different from those of other machine learning systems, specifically the back-propagation neural networks and the C4.5 symbolic induction algorithm. With its unique features, supervised ART has offered an interesting alternative approach to many real-world problems. Two applications are described below.

In personalized information systems, supervised ART systems can be used to model users' profile so that only the information most relevant to a user is identified and presented [17]. Each user profile is represented by a set of recognition categories, each associating a set of conjunctive features of a piece of information to a relevance factor. As the network structure is compatible with rule-based knowledge, user-defined rules can be readily translated into the recognition categories of a supervised ART system. In addition, subsequent user feedback on individual pieces of information can be used to refine the network. Through the refinement process, the network learns interest terms that are not explicitly mentioned by the user. As both user-defined and system-learned knowledge are represented in a single system, any inherent conflict or inconsistency can be detected and resolved readily.

Another potential domain is that of knowledge discovery and interpretation. Traditional data mining tools do not incorporate users' domain knowledge in the knowledge discovery process. As a result, the discovered knowledge can be very different from the users' perspectives and difficult to interpret. Supervised ART, on the other hand, provides a mechanism to incorporate users' knowledge. By building upon a user's prior knowledge, the final result is expected to be more interpretable to the user.

Acknowledgments

This chapter is based on the author's dissertation thesis submitted to the Department of Cognitive and Neural Systems, Boston University. The author would like to acknowledge Gail A. Carpenter, Stephen Grossberg, and Michael Cohen for their guidance and support. Thanks also go to Victor Chew who gave valuable comments on a previous version of the manuscript.

References

[1] Y. R. Asfour, G. A. Carpenter, S. Grossberg, and G. W. Lesher. Fusion ARTMAP: A neural network architecture for multi-channel data fusion and classification. In *World Congress on Neural Network, Portland, OR*, volume II, pages 210–215. Hillsdale, NJ: Lawrence Erlbaum Associates, July 1993.

[2] G. A. Carpenter and S. Grossberg. A massively parallel architecture for a self-organizing neural pattern recognition machine. *Computer*

Vision, Graphics, and Image Processing, 37:54–115, 1987.

[3] G. A. Carpenter, S. Grossberg, N. Markuzon, J. H. Reynolds, and D. B. Rosen. Fuzzy ARTMAP: A neural network architecture for incremental supervised learning of analog multidimensional maps. *IEEE Transactions on Neural Networks*, 3:698–713, 1992.

[4] G. A. Carpenter, S. Grossberg, and J. H. Reynolds. ARTMAP: Supervised real time learning and classification by a self-organizing neural network. *Neural Networks*, 4:565–588, 1991.

[5] G. A. Carpenter, S. Grossberg, and D. B. Rosen. Fuzzy ART: Fast stable learning and categorization of analog patterns by an adaptive resonance system. *Neural Networks*, 4:759–771, 1991.

[6] G. A. Carpenter and A.-H. Tan. Rule extraction: From neural architecture to symbolic representation. *Connection Science*, 7(1):3–27, 1995.

[7] R. Farber, A. Lapedes, and K Sirotkin. Determination of eucaryotic protein coding regions using neural networks and information theory. *Journal of Molecular Biology*, 226:471–479, 1992.

[8] L. M. Fu and L. C. Fu. Mapping rule-based knowledge into neural architecture. *Knowledge-Based Systems*, 3:48–56, 1990.

[9] R. C. Lacher, S. I. Hruska, and D. C. Kuncicky. Backpropagation learning in expert networks. *IEEE Transactions on Neural Networks*, 3:62–72, 1992.

[10] A. Lapedes, C. Barnes, C. Burks, R. Farber, and K Sirotkin. Application of neural networks and other machine learning algorithms to dna sequence analysis. In *Computers and DNA, SFI Studies in the sciences of complexity, vol VII*, pages 265–281. Reading, MA: Addison-Wesley, 1990.

[11] P. M. Murphy and D. W. Aha. UCI repository of machine learning databases [machine-readable data repository]. Irvine, CA: University of California, Department of Information and Computer Science, 1992.

[12] M. O'Neill. Escherichia coli promoters: I. Consensus as it relates to spacing class, specificity, repeat substructure, and three dimensional organization. *Journal of Biological Chemistry*, 264:5522–5530, 1989.

[13] M. O'Neill. Escherichia coli promoters: II. A spacing class-dependent promoter search protocol. *Journal of Biological Chemistry*, 264:5531–5534, 1989.

[14] J. R. Quinlan. Induction of decision trees. *Machine Learning*, 1:81–106, 1986.

[15] A.-H. Tan. Adaptive Resonance Associative Map. *Neural Networks*, 8(3):437–446, 1995.

[16] A.-H. Tan. Cascade ARTMAP: Integrating neural computation and symbolic knowledge processing. *IEEE Transactions on Neural Networks*, 8(2):237–250, 1997.

[17] A.-H. Tan and C. Teo. Learning user profiles for personalized information dissemination. In *Proceedings, 1998 IEEE International Joint Conference on Neural Networks, Alaska*, pages 183–188, 1998.

[18] G. G. Towell and J. W. Shavlik. Interpretation of artificial neural networks: Mapping knowledge-based neural networks into rules. In *Advances in Neural Information Processing Systems 4*, pages 977–984. San Mateo, CA: Morgan Kaufmann, 1992.

[19] G. G. Towell and J. W. Shavlik. Extracting refined rules from knowledge-based neural networks. *Machine Learning*, 13:71–101, 1993.

[20] G. G. Towell, J. W. Shavlik, and M. O. Noordewier. Refinement of approximately correct domain theories by knowledge-based neural networks. In *Proceedings, 8th National Conference on AI, Boston, MA*, pages 861–866. AAAI Press/The MIT Press, 1990.

[21] V. Tresp, J. Hollatz, and S. Ahmad. Network structuring and training using rule-based knowledge. In *Advances in Neural Information Processing Systems 5*, pages 977–984. San Mateo, CA: Morgan Kaufmann, 1993.

[22] M. S. Waterman. *Mathematical Methods for DNA sequences.* Boca Raton, FL: CRC Press, 1989.

CHAPTER 5

INCREMENTAL LEARNING AND OPTIMIZATION OF HIERARCHICAL CLUSTERINGS WITH ART-BASED MODULAR NETWORKS

G. Bartfai and **R. White**
School of Mathematical and Computing Sciences
Victoria University of Wellington
New Zealand
{guszti,rwhite}@mcs.vuw.ac.nz

This chapter introduces HART-S, a modular neural network that can incrementally learn stable hierarchical clusterings of arbitrary sequences of input patterns by self-organisation. The network is a cascade of Adaptive Resonance Theory (ART) modules, in which each module learns to cluster the *differences* between the input pattern and the selected category prototype at the previous module. Input patterns are first classified into a few broad categories, and successive ART modules find increasingly specific categories until a threshold is reached, the level of which can be controlled by a global parameter called "resolution". The network thus essentially implements a divisive (or splitting) hierarchical clustering algorithm: hence the name HART-S (for "Hierarchical ART with Splitting"). HART-S is also compared and contrasted to HART-J (for "Hierarchical ART with Joining"), another variant that was proposed earlier by the first author. The network dynamics are specified and some useful properties of both networks are given and then proven. Experiments were carried out on benchmark datasets to demonstrate the representational and learning capabilities of both networks and to compare the developed clusterings with those of two classical methods and a conceptual clustering algorithm. Two optimisation methods for the HART-S network are also introduced. The aims are to reduce the complexity of trained networks and "clean up" the category prototypes during the learning process while maintaining the useful properties of hierarchical ART networks like fast and stable learning, and the ability to build category hierarchies incrementally. The experimental results demonstrate a significant reduction in category complexity as well as some improvement on a range of other metrics at a cost of varying amounts of additional training time. We suggest that scheduling the optimisation steps may be crucial in achieving an optimal trade-off. A brief survey of related neural network models is also provided.

1 Introduction

The combination of neural networks has gained increasing attention in recent years. Although some monolithic "black-box like" networks, like the multi-layer perceptron, have been shown to be universal function approximators and used in many practical applications, they have some well-known disadvantages, the most severe of those is intractable training time. One way to alleviate this problem is to divide the task into smaller ones and train the individual – and proportionally smaller – networks on each sub-task separately. This modular approach appears to be supported by findings in brain study which suggest that the human brain is not just one big mass of neurons. For example, neurons of the neocortex, where most high-level cognitive processes occur, are of few types, and their connections are organised in relatively few schemes or basic circuits [40]. The primate visual system has also been found to be highly modularised with extensively interconnected subdivisions performing linear and non-linear filtering, object recognition and so on [42]. The arguments for decomposition have engineering, complexity, and psychological aspects as well [26]. There are also a number of theoretical advantages: coding can be sparse; scalability of solutions; modularity of solution leading to self-similar fractal-like architectures [23]. Consequently, modular neural network models have been proposed, (see for example [25], [36]) which possess distinctive advantages over monolithic ones like faster training [33] and better overall performance [30].

Besides smaller computational costs, modular networks can also be used to solve more complex tasks that simply cannot be done (i.e. represented and learned) by a single network module. One such area is unsupervised clustering, which is the main focus of this chapter. The ability to learn about the environment without a teacher has long been considered an important characteristic of intelligent systems. Unsupervised learning can be found both at sensory-level of mammals, and at higher, cognitive levels of humans. Therefore, it has been an important topic in neural network research (e.g. [31], [38], [39]) in recent years. Unsupervised learning networks typically perform dimensionality reduction or pattern clustering. In the latter case, upon presentation of an input pattern, the node whose connection weight vector is the closest to the current input vector (in some distance metrics) will become the only active node (winner-take-all competition), and will be allowed to learn, i.e., modify its connection weights. After repeated exposure to the input environment, the network will store a prototypical element of each category, or class, in its connections weights. The number and size of classes the network finds on a given training set depends on the number of output nodes. The more neurons the network has, the larger number of more specific categories it will find. Because of its "winner-take-all" characteristics, the network will only be able to find *one* category for a given input pattern, regardless of how large the network is. However, environments are often more complex and may exhibit a hierarchical structure (e.g. taxonomy of animals), which humans can learn without apparent difficulty. It is therefore reasonable to assume that network circuitry which performs hierarchical clustering is present in the brain. An example for such a network is discussed in [1], where it was suggested that the paleocortex performs hierarchical clustering. The authors also hypothesise that "networks and circuitry of similar design in other brain regions contain computational elements sufficient to construct perceptual hierarchies for use in recognizing environmental cues", and suggest the "possibility that this biologically generated mechanism for hierarchical clustering may be a routine part of perceptual recognition memory behavior in animals and humans".

It is, therefore, worthwhile investigating neural network architectures that can learn class hierarchies. Networks that contain multiple competitive layers appear to be suitable candidates. For example, Rumelhart et al. [39] suggest that a hierarchical structure can be formed if a network has several "inhibitory clusters" in each layer that work in parallel, and each of them receives input from the output of the previous layer only (which is a vector with the winners in each active cluster). This method was shown to have some advantages on an artificial example, but it is not clear – due to its inherently probabilistic nature – whether it can be used successfully in other, possibly more complex tasks. Standard competitive learning networks also have several drawbacks like long training time, finding and getting stuck in suboptimal solutions, and instability against changes in the environment.

An alternative model that is based on solid neurophysiological and cognitive principles, and also overcomes the above problems, is the Adaptive Resonance Theory (ART) neural network [12]. They are more suitable for the systematic construction of modular networks because of their well-defined interfaces as well as features that most other networks lack. In specific, ART networks have the ability to create new output nodes (i.e. categories) dynamically, and do not suffer from the problem of forgetting previously learned categories if the environment changes. In other words, it provides a solution to the stability-plasticity dilemma [16]. A neural network with similar functionalities (CALM, for "CAtegorizing and Learning Module") was proposed as a computational primitive in more complex, modular networks [23]. Despite their useful properties, however, ART networks by themselves can only learn input categories at a given level of specificity (i.e. partition the input space), which depends on a global parameter called *vigilance*.

The central idea of this chapter is to combine ART networks in a systematic way into a modular architecture which is capable of learning *class hierarchies* from arbitrary sequences of input patters, while retaining the main properties of single ART networks, namely autonomous, fast and stable learning by self-organisation. In specific, we introduce a new ART-based modular network that develops cluster hierarchies by first creating large, general classes, which are subsequently refined in lower layers by splitting up each super-class into sub-classes (by clustering features that are different from the super-class). The biological plausibility of such mechanism was suggested in [1]. We call this "splitting" – a term used in statistical clustering after [24] – hierarchical ART network, or HART-S (for "Hierarchical ART with Splitting"). In particular, we contrast its functionality and properties to those of the "joining" hierarchical ART network (HART-J), which was proposed independently by Bartfai [5] and Ishihara [28] under the names *HART* and *arboART*, respectively. (We discuss some related works in more detail in Section 6.) The HART-J network has several layers of ART modules where each layer learns to cluster the *prototypes* the previous layer has developed. The network therefore builds successively larger classes in upper layers by joining the lower level, more specific, ones.

Section 2 describes the ART network and summarises its properties. The hierarchical ART models are introduced in Section 3. The joining hierarchical ART model (HART-J) and its properties are discussed in Sections 3.1 and 3.1.2, respectively. The splitting hierarchical ART model (HART-S) and its properties are discussed in Sections 3.2 and 3.2.2, respectively. The experiments that were carried out on a machine learning database are shown in Section 4. The two optimisation methods for the HART-S network are introduced in Section 5, where experimental results are also presented. The results and related network models are discussed in Section 6. Finally, conclusions are drawn in Section 7.

2 The ART Network

This section describes the ART architecture at a level of detail that is necessary for understanding the rest of the chapter. Unless otherwise stated, we refer to ART1, the binary version of ART [12][1].

Adaptive Resonance Theory (ART) architectures are neural networks that develop stable recognition codes by self-organisation in response to arbitrary sequences of input patterns. They were designed to solve the "stability-plasticity dilemma" that every intelligent machine learning system has to face: how to keep learning from new events without forgetting previously learned information.

An ART network is built up of three layers: the *input* layer (F0), the *comparison* layer (F1) and the *recognition* layer (F2) with M, M and N neurons, respectively (see Figure 1). The input layer stores the input pattern, and each neuron in the input layer is connected to its corresponding node in the comparison layer via one-to-one, non-modifiable links. Nodes in

[1]There are ART networks that accept both continuous and binary inputs [11], [15].

the F2 layer represent input categories. The F1 and F2 layers interact with each other through weighted bottom-up and top-down connections that are modified when the network learns. The G1, G2 and R signals control the operation of the network: the non-specific *gain control* signals G1 and G2 are used to enable/disable all the nodes in the F1 and F2 layers, respectively. The main role of G1 is to implement template matching in the F1 layer (see equation (2) below, where use of the *reset signal* R is also described). The learning process of the network can be described as follows.

At each presentation of a non-zero binary input pattern $\mathbf{x}$ ($x_i \in \{0,1\}$, $i = 1, 2, \ldots, M$), the network attempts to classify it into one of its existing categories based on its similarity to the stored prototype of each category node. More precisely, for each node i in the F2 layer, the bottom-up activation T_j is calculated, which can be expressed as

$$T_j = \frac{|\mathbf{w}_j \cap \mathbf{x}|}{\beta + |\mathbf{w}_j|} \quad j = 1, \ldots, N \tag{1}$$

where $|\cdot|$ is the norm operator ($|\mathbf{x}| \equiv \sum_{i=1}^{M} x_i$), $\mathbf{w}_j$ is the (binary) weight vector [2] (or prototype) of category j, and $\beta > 0$ is the *choice parameter* [14]. Then the F2 node J that has the highest bottom-up activation, i.e., $T_J = \max\{T_j | j = 1, \ldots, N\}$, is selected (winner-take-all competition). The weight vector of the winning node ($\mathbf{w}_J$) will then be compared to the current input at the comparison layer. If they are similar enough, i.e., they satisfy the

$$\frac{|\mathbf{w}_J \cap \mathbf{x}|}{|\mathbf{x}|} \geq \rho \tag{2}$$

matching condition, where ρ is a system parameter called *vigilance* ($\rho \in [0,1]$), F2 node J will capture the current input and the network learns by modifying $\mathbf{w}_J$:

$$\mathbf{w}_J^{new} = \eta(\mathbf{w}_J^{old} \cap \mathbf{x}) + (1 - \eta)\mathbf{w}_J^{old} \tag{3}$$

[2] Since the bottom-up weights are essentially normalised versions of the corresponding (binary) top-down ones, subsuming them into a single weight vector – and modifying the network equations accordingly – is a commonly used simplification, which we shall adopt throughout the chapter. This also assumes the network is in fast learning mode ($\eta = 1$, see below).

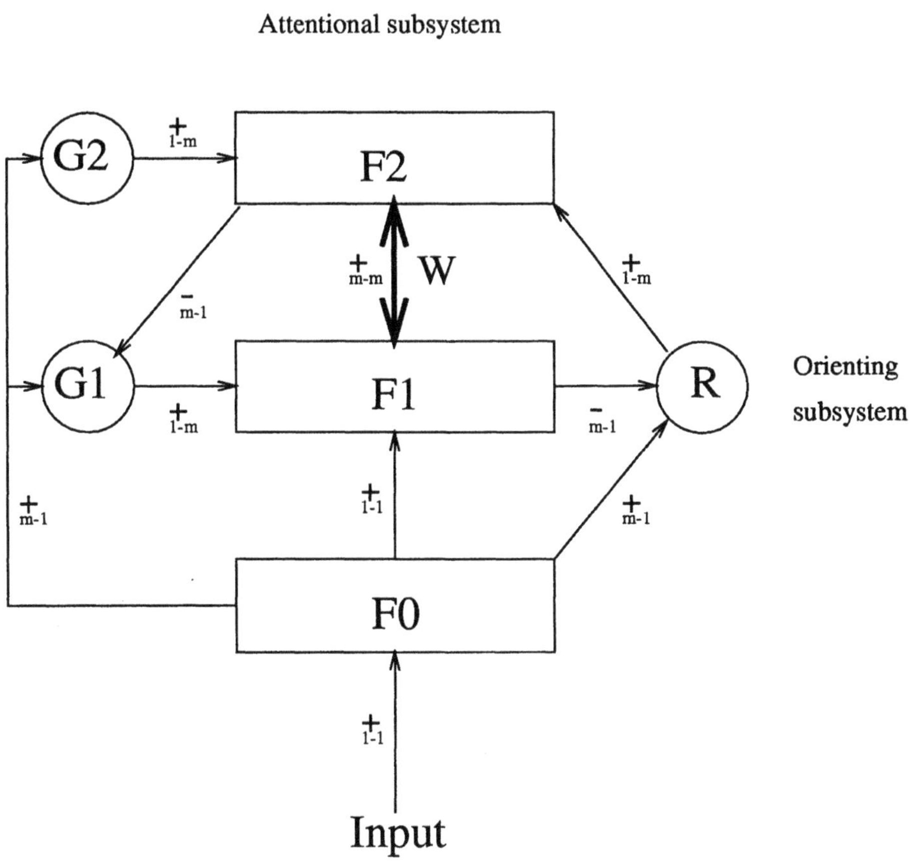

Figure 1. Architecture of the ART network. Thick arrows denote modifiable connections. The "+" and "−" signs represent excitatory and inhibitory connections, respectively. The symbols $\{1, m\} - \{1, m\}$ represent {one,many}-to-{one,many} connections. Winning F2 category nodes are selected by the *attentional* subsystem. Category search is controlled by the *orienting* subsystem. If the degree of category match at the F1 layer is lower than the vigilance level ρ, the non-specific reset signal R will be triggered, which will deactivate the winning F2 node for the period of presentation of the current input.

where η is the learning rate ($\eta \in [0, 1]$). All other weights in the network remain unchanged.

If, however, the stored prototype $\mathbf{w}_J$ does not match the input sufficiently, i.e., condition (2) is not met, the winning F2 node will be reset (by activating the reset signal R in Figure 1) for the period of presentation of the current input. Then another F2 node (or category) is selected with the highest T_j whose prototype will be matched against the input, and so on. This "hypothesis-testing" cycle is repeated until the network either finds a stored category whose prototype matches the input well enough, or allocates a new F2 node. Then learning takes place according to (3).

It is important to note that once a category is found, the comparison layer (F1) holds $\mathbf{w}_J \cap \mathbf{x}$ until the current input is removed. It can be shown that after an initial period of self-stabilisation, the network will directly (i.e. without search) access the prototype of one of the categories it has found in a given training set.

Note also, however, that, as a consequence of its stability-plasticity property, the network is capable of learning "on-line", i.e., refining its learned categories in response to a stream of input patterns (as opposed to being trained "off-line" on a finite training set).

The set of attribute values described by ART cluster prototypes correspond to the intersection of all those attribute values found in instances of that cluster. In terms of the clustering algorithm this is not as restrictive as it first seems. A common technique used with ART is the complement coding of input patterns [14]. The result of using this together with 1-of-n encoded attribute values is that one half of a cluster prototype corresponds to the intersection of the attribute values that all instances in that cluster do *not* have, or from another angle, the union of the attribute values present in that cluster. In total then a cluster prototype specifies a range of values for each attribute and has an effect which, combined with Fuzzy ART [15], adds significantly to the clustering capability of the mechanism.

The number of developed categories can be controlled by setting ρ: the higher the vigilance level, the larger number of more specific categories will be created. (If $\rho = 1$, the network will create a new category for every unique input.)

In order to get two or more "views" of the training set with the same network, one can consider choosing low vigilance first to get an overall view, and then raise it at a later phase during training to gain a more detailed view as well. This, however, will result in the network not being able to access the overall view (i.e. coarser categorisation) anymore, but retaining only the more specific categories, even if vigilance is lowered again later. This is because only one category is selected for a given input, and if two categories ($\mathbf{w}_l$, $\mathbf{w}_h$) were found to match the current input perfectly, i.e.,

$$|\mathbf{w}_l \cap \mathbf{x}| = |\mathbf{w}_l|, \quad \text{and} \quad |\mathbf{w}_h \cap \mathbf{x}| = |\mathbf{w}_h| \,,$$

the more specific one (i.e. of the larger norm) will be preferred due to (1). There is no relationship between any pairs of category prototypes except that they are in the same network and compete with each other. The network, therefore, with its single layer of category nodes, is not capable of representing (and thus learning) a hierarchy of classes. This motivated us to design networks with multiple ART modules.

3 Hierarchical ART Networks

This section discusses two ART-based architectures that can perform hierarchical clustering by incremental learning. Both of them are essentially a cascade of ART modules with the input pattern entering at one end. Information processed by an ART module is passed on to the next one (see Figure 2). At this level, it is basically the kind of information that is passed on that distinguishes the two models.

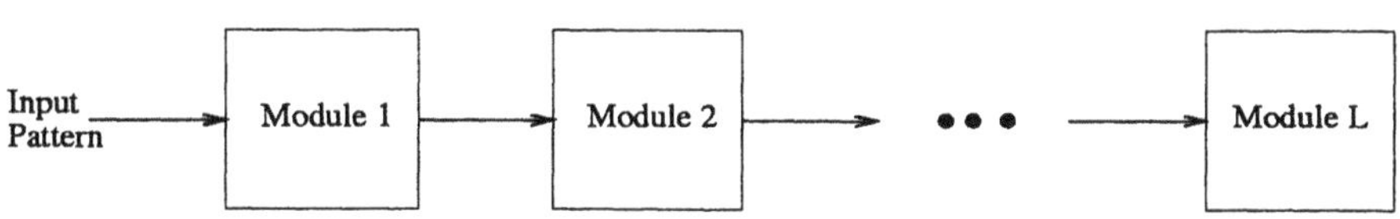

Figure 2. Overall view of the architecture of a Hierarchical ART network. Each module is an ART network.

The "joining" HART network (HART-J) first clusters input patterns into (potentially many) specific categories (with high vigilance level in the first module), and then successive layers cluster the *prototypes* that are developed at the previous layer. This architecture was first presented in [5]. Here we describe its main features as well as some further properties, some of which are particularly useful when comparing them to those of the "splitting" HART network. The HART-S network first creates a few, typically broad, clusters. Then successive layers learn to cluster the *differences* between a prototype and the input vector at the previous layer.

3.1 The Joining Hierarchical ART Network (HART-J)

This model is based on the idea of starting with the most specific clustering of the input domain and progressively merging such clusters into more general ones. Only the similarity between an input to an ART module and the prototype of the category of that input is propagated to the next ART module. The input signal to the first module is the similarity between the input pattern and its most specific categorisation (itself), so is equivalent to the input pattern itself. This similarity is exactly the content of the F1 layer during resonance, so the structure of HART-J reflects this with two connections between successive ART modules (Figure 3). One connection propagates inhibitory signals from global reset to the gain controls of a second module so that it cannot become active until a resonant state is obtained in the first [3]. The other connection propagates the contents of the F1 layer of the first module to the F0 layer of the second.

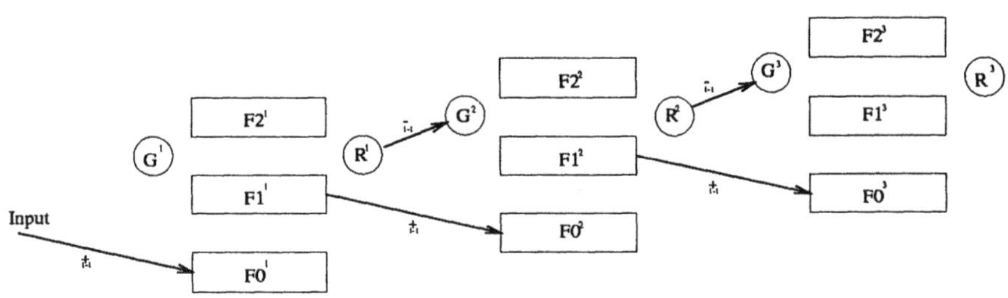

Figure 3. Architecture of a three-layer HART-J network. Connections within the ART modules are not shown. The G1 and G2 ART signals are represented with one G signal for each module ($G^i, i = 1,2,3$). The meanings of the "+", "−" and "1 − 1" symbols are explained at Figure 1.

More formally, the HART-J network is described by the following parameter tuple:

$$\mathcal{HJ} = \{L, \mathbf{M}, \mathbf{N}, \boldsymbol{\beta}, \boldsymbol{\rho}, \boldsymbol{\eta}\}$$

where the meaning of each parameter is as follows:

[3]In real-time implementations, this will involve specifying the dynamics of these connections so as to filter out undesirable "spikes" of the reset signal during category search.

L : the number of ART modules used
$\mathbf{M} = (M_1, \ldots, M_L)$: the size of each F0 and F1 layer
$\mathbf{N} = (N_1, \ldots, N_L)$: the size of each F2 layer
$\boldsymbol{\beta} = (\beta_1, \ldots, \beta_L)$: the choice parameter of each ART module ($0 < \beta_k$, typically $\beta_k \ll 1$)
$\boldsymbol{\rho} = (\rho_1, \ldots, \rho_L)$: the vigilance level of each ART module, $\rho_k \in [0, 1]$
$\boldsymbol{\eta} = (\eta_1, \ldots, \eta_L)$: the learning rate of each ART module, $\eta_k \in [0, 1]$

with the architectural constraint of $M_k = M$ for $k = 1, \ldots, L$. For further simplicity, however, we will also assume $N_k = N$, $\beta_k = \beta$ and $\eta_k = \eta = 1$ (i.e., fast learning) for $k = 1, \ldots, L$. So it is only the ρ_k vigilance levels that are allowed to be different in each ART module.

From Figure 3 and noting the contents of the F1 layer in resonance (see Section 2), the local input to ART module k can be expressed as

$$\mathbf{x}_k = \mathbf{x}_{k-1} \cap \mathbf{w}_{k-1:J}, \quad k = 2, \ldots, L \tag{4}$$

where $_{k-1:J}$ is the index of the winning F2 node in ART module $k - 1$. The prototype (or weight vector) of the winning category $_{k:J}$ in layer k after training (see equation (3) if $\eta = 1$) becomes

$$\mathbf{w}_{k:J} = \mathbf{x}_k \cap \mathbf{w}_{k:J}, \quad k = 1, \ldots, L\,. \tag{5}$$

Applying (4) and (5) to layers $k-1, k-2, \ldots, 2$ recursively (and ignoring the indices of winning F2 nodes for simplicity), we get

$$\mathbf{w}_k = \mathbf{x}_1 \cap \mathbf{w}_1 \cap \ldots \cap \mathbf{w}_k, \quad k = 1, \ldots, L \tag{6}$$

where $\mathbf{x}_1$ is the *local* input to the first ART module, which is also the *global* input $\mathbf{X}$ to the entire HART-J network. Equation (6) can then be written as

$$\mathbf{w}_k = \mathbf{X} \cap \mathbf{w}_1 \cap \ldots \cap \mathbf{w}_k = \mathbf{X} \cap \mathbf{W}_k, \quad k = 1, \ldots, L \tag{7}$$

where $\mathbf{W}_k$ (or more precisely $\mathbf{W}_{k:J}$) is the *global* (or *hierarchical*) prototype of level k, i.e., the prototype of category J in level k *with respect to the global* input $\mathbf{X}$. From (4) and (5), it can be seen that

$$\mathbf{w}_k = \mathbf{w}_{k-1} \cap \mathbf{w}_k\ , \tag{8}$$

which shows that the network develops more general categories at successively higher levels. It also follows from (7) and (8) that

$$\mathbf{W}_k = \mathbf{w}_k, \quad k = 1, \ldots, L \ . \tag{9}$$

Since the HART-J network develops more general hierarchical categories, it is instructive to ask what the *maximum* degree of specificity any level k category can have given a sequence of training patterns of size S and (local) vigilance levels $\rho_1, \ldots, \rho_k$. In a single ART network, this is equivalent to asking what the minimum level of generality of any category will be, which is simply given by the (only) vigilance level ρ since $|\mathbf{w}_J| \geq \rho|\mathbf{x}|(= \rho S)$ [4] for any input $\mathbf{x}$ and winning node J. For a given level in a HART-J network, however, the two limits (i.e. minimum level of generality and maximum level of specificity) are separated since the local inputs to each level will vary due to the generalisation done by previous modules. So in a HART-J network, there will be a region of *effective vigilance* at level k, which provides bounds of the specificity that level can have with respect to the global input $\mathbf{X}$. A discussion on the lower bound of this region can be found in [5]. Here we provide the upper bound ($\Omega_{k:max}$), which is of more use for practical considerations. (For example, it allows us to put a bound on the number of ART modules needed in an application by guaranteeing a given level of specificity that level k cannot exceed.) More specifically, in each level we look for the smallest value z such that

$$\frac{|\mathbf{w}_k|}{|\mathbf{x}_k|} < \rho_k \rightarrow \frac{|\mathbf{W}_k|}{|\mathbf{X}|} < z \ . \tag{10}$$

The value of $\Omega_{k:max}$ is given with proof in Section 3.1.2.

3.1.1 Dynamics of HART-J

The dynamics of the HART-J network are closely linked to those of component ART modules, and are described as follows.

For $1 \leq i \leq M, 1 \leq j \leq N$, and $1 \leq k \leq L$

[4] Inputs here are assumed to be of constant norm only for the sake of simplicity. The discussion will still hold, with minor modifications, if the inputs are not normalised.

- *Initialisation*

  The initial values of all the connection weights are

$$w_{k:ji} = 1$$

- *Layer activity vectors*

  The $F0_k$ layer:

$$\mathbf{x}_k = \mathbf{y1}_{k-1}$$

  The $F1_k$ layer:

$$\mathbf{y1}_k = \begin{cases} \mathbf{x}_k \cap \mathbf{w}_{k:j} & \text{when } F2_k \text{ node } j \text{ is active and module } k-1 \\ & \text{is in resonance} \\ \mathbf{x}_k & \text{otherwise} \end{cases}$$

  The $F2_k$ layer:

$$y2_{k:j} = \begin{cases} 1 & \text{when } F2_k \text{ node } j \text{ is active and module } k-1 \text{ is in} \\ & \text{resonance} \\ 0 & \text{otherwise} \end{cases}$$

  Whenever an input is presented, and module $k-1$ is in resonance, all $F2_k$ nodes are enabled and an initial category choice is made.

- *Category Choice*

  For any module k, if module $k-1$ is in resonance, choose some j such that $F2_k$ node j is enabled and

$$\frac{|\mathbf{w}_{k:j} \cap \mathbf{x}_k|}{\beta + |\mathbf{w}_{k:j}|}$$

  is maximal, with the result that $F2_k$ node j becomes active.

- *Category Match*

  In any module k a local category matches a local input when $F2_k$ node j is active, module $k-1$ is in resonance and

$$\frac{|\mathbf{w}_{k:j} \cap \mathbf{x}_k|}{|\mathbf{x}_k|} \geq \rho_k \tag{11}$$

  If the match is unsuccessful then $F2_k$ node j is disabled and another category choice is made.

  If the match is successful then learning takes place and the system is stable until the input is removed.

- *Learning*

  When $F2_k$ node j is active and (11) holds·

$$\mathbf{w}_{k:j}^{new} = \eta(\mathbf{w}_{k:j}^{old} \cap \mathbf{x}_k) + (1 - \eta)\mathbf{w}_{k:j}^{old}$$

3.1.2 Properties of HART-J

This section describes the main properties of the HART-J network.

1. *Stability of Learning.*

   After a finite number of input presentations of an input set, all the individual ART modules will stabilise if the input is normalised. The following argument explains why this is so.

   Let the norm of the input set be S. Then the smallest size of a local input to module $k > 1$ is $S \prod_{i=1}^{k-1} \rho_i$. The learning of any ART module with normalised inputs will stabilise because such normalisation is a limit on the smallest size of an input [15]. This lower bound on the size of inputs has the same effect.

2. *Order of Presentation.*

   The order of presentation affects the clustering. This follows naturally from the properties of the individual ART modules [37].

3. *Direct Access.*

   Once learning has stabilised direct access to category prototypes is assured for each ART module. Given an input presentation, each module must wait for resonance in previous modules but no search is required.

4. *An Upper Bound on Effective Vigilance.*

$$\Omega_{k:max} = \rho_k$$

   Proof

   In this model $\mathbf{w}_k = \mathbf{W}_k$.

The worst case is when $\mathbf{x}_k$ is as big as possible, in fact in this model it could be up to the same size as $\mathbf{X}$.

So the least z such that

$$\frac{|\mathbf{W}_k|}{|\mathbf{X}|} < \rho_k \rightarrow \frac{|\mathbf{W}_k|}{|\mathbf{X}|} < z$$

for any $\mathbf{X}$ and any clustering is ρ_k. □

3.2 The Splitting Hierarchical ART Network (HART-S)

This new network model we are proposing is based on the idea of forming the most general level of clusters first and then successively splitting those clusters into more specific ones. Only the difference between an input to a particular ART module and the prototype of the category of that input is propagated to the next ART module. The input signal to the first module is the difference between the input pattern and the most general category, so is equivalent to the input pattern itself. Only when such differences are greater than a size determined by a network parameter called resolution (ϵ) is the next ART module activated.

More formally, the HART-S network is described by the following parameter tuple:

$$\mathcal{HS} = \{L, \mathbf{M}, \mathbf{N}, \boldsymbol{\beta}, \boldsymbol{\rho}, \epsilon, \boldsymbol{\eta}\}$$

where the meaning of each parameter is as follows:

L :	the number of ART modules used
$\mathbf{M} = (M_1, \ldots, M_L)$:	the size of each F0 and F1 layer
$\mathbf{N} = (N_1, \ldots, N_L)$:	the size of each F2 layer
$\boldsymbol{\beta} = (\beta_1, \ldots, \beta_L)$:	the choice parameter of each ART module ($0 < \beta_k$, typically $\beta_k \ll 1$)
$\boldsymbol{\rho} = (\rho_1, \ldots, \rho_L)$:	the vigilance level of each ART module, $\rho_k \in [0, 1]$
ϵ :	resolution parameter, $\epsilon \in [0, 1]$
$\boldsymbol{\eta} = (\eta_1, \ldots, \eta_L)$:	the learning rate of each ART module, $\eta_k \in [0, 1]$

with the architectural constraint of $M_k = M$ for $k = 1, \ldots, L$. For further simplicity, however, we will also assume $N_k = N$, $\beta_k = \beta$ and $\eta_k = \eta = 1$ (i.e., fast learning) for $k = 1, \ldots, L$. So it is only the ρ_k vigilance levels that are allowed to be different in each ART module.

The architecture of a three-layer HART-S network is shown in Figure 4. There are four sets of connections between each successive module in a HART-S network (Figure 4). An adaptive associative connection (V) called hierarchical memory between the F2 layers of successive ART modules provides the links between any cluster and its immediate sub-clusters. When a category is active in the first of a pair of modules, only those categories in the second module that can be sub-categories of the first are allowed to become active. These links are important in this network (which are not needed in HART-J) so the correct hierarchies can be maintained, since the individual ART modules cluster *differences* only. The other inter-ART connections are explained in Figure 4.

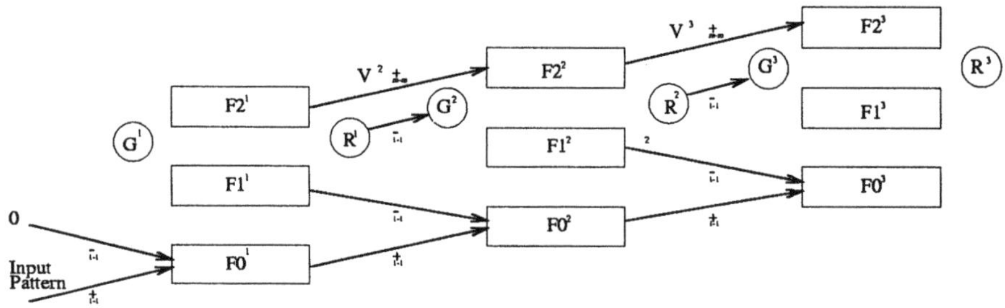

Figure 4. Architecture of a three-layer HART-S network. Only the connections between ART modules are shown. V^2 and V^3 are (many-to-many) associative links between F2 nodes of neighbouring modules. The $R^i \rightarrow G^{i+1}$ links have the same purpose as in HART-J. The $F0^i \rightarrow F0^{i+1}$ and $F1^i \rightarrow F0^{i+1}$ one-to-one links are used to propagate the difference between the current input and the selected category prototype to the next layer.

The local input to level k is

$$\mathbf{x}_k = \mathbf{x}_{k-1} - \mathbf{x}_{k-1} \cap \mathbf{w}_{k-1:J}, \quad k = 2, \ldots, L \tag{12}$$

where $_{k-1:J}$ is the index of the winning F2 node in module $k-1$. Since both the inputs and prototypes are binary, (12) can be written as

$$\mathbf{x}_k = \mathbf{x}_{k-1} \cap \overline{\mathbf{w}}_{k-1:J}, \quad k = 2, \ldots, L \tag{13}$$

where $\bar{\cdot}$ denotes the Boolean negation operator. Applying (5) and (13) to layers $k, k-1, \ldots, 2$ recursively (and ignoring the indices of winning F2 nodes for simplicity), for the local input to layer $k+1$ we get

$$\mathbf{x}_{k+1} = \mathbf{x}_1 \cap \overline{\mathbf{w}}_1 \cap \ldots \cap \overline{\mathbf{w}}_k = \mathbf{X} \cap (\overline{\mathbf{w}_1 \cup \ldots \cup \mathbf{w}_k}) = \mathbf{X} \cap \overline{\mathbf{W}}_k, \quad k = 1, \ldots, L \tag{14}$$

where $\mathbf{W}_k$ (or more precisely $\mathbf{W}_{k:J}$) is the *global prototype* of category J at level k:

$$\mathbf{W}_k = \mathbf{w}_1 \cup \ldots \cup \mathbf{w}_k, \quad k = 1, \ldots, L \tag{15}$$

Note that

$$\mathbf{w}_i \cap \mathbf{w}_j = \emptyset, \quad i \neq j \in 1, \ldots, k\ ,$$

which follows from the property that $\mathbf{w}_i = \mathbf{x}_i \cap \mathbf{w}_i$ and $\mathbf{w}_j = \mathbf{x}_j \cap \mathbf{w}_j$ and that $\mathbf{x}_j \cap \mathbf{w}_j = \mathbf{x}_i \cap \overline{\mathbf{w}}_i \cap \overline{\mathbf{w}}_{j-1} \cap \ldots \cap \mathbf{w}_j$ assuming $i < j$.

A distinctive feature of the HART-S network is that the local input $\mathbf{x}_k$ will *not* activate module k unless $|\mathbf{x}_k| > \epsilon S$. Thus ϵ (or resolution) controls the amount of sub-clustering that can occur.

Since the HART-S network develops more specific hierarchical (or global) categories in successive levels, it is useful to ask what the *minimum* degree of specificity any level k category can have given a sequence of training patterns of size S and (local) vigilance levels $\rho_1, \ldots, \rho_k$. For similar reasons to those in HART-J, a single *effective vigilance* level cannot be determined from the individual vigilance levels. Instead, there will be a region, whose lower bound ($\Omega_{k:min}$) will be of interest to us. More specifically, in each level we look for the smallest value z such that

$$\frac{|\mathbf{w}_k|}{|\mathbf{x}_k|} \geq \rho_k \rightarrow \frac{|\mathbf{W}_k|}{|\mathbf{X}|} \geq z\ . \tag{16}$$

The value of $\Omega_{k:min}$ is given with proof in Section 3.2.2.

Note that the effective vigilance for different clusters in a given module of a HART-S network may be quite different (and this precludes using effective vigilance as a global parameter in some way). It is an important part of the mechanism since it allows the information found in the input environment to determine the shape of the resulting hierarchy.

3.2.1 Dynamics of HART-S

The dynamics of the HART-S network can be described by those of the component ART modules and the data/memory connections between them as follows.

For $1 \leq i \leq M, 1 \leq j, h \leq N$, and $1 \leq k \leq L$

- *Initialisation*

  The initial values of all connection weights are:

$$v_{k:jh} = 1$$

$$w_{k:ji} = 1$$

- *Layer activity vectors*

  The $F0_k$ layer:

$$\mathbf{x}_k = \mathbf{x}_{k-1} - \mathbf{y1}_{k-1}$$

  with $\mathbf{x}_1 \equiv \mathbf{x}_0 \equiv \mathbf{X}$ and $y1_{0:i} = 0$.

  The $F1_k$ layer:

$$\mathbf{y1}_k = \begin{cases} \mathbf{x}_k \cap \mathbf{w}_{k:j} & \text{when } F2_k \text{ node } j \text{ is active and} \\ & \text{module } k-1 \text{ is in resonance} \\ \mathbf{x}_k & \text{otherwise} \end{cases}$$

  The $F2_k$ layer:

$$y2_{k:j} = \begin{cases} 1 & \text{when } F2_k \text{ node } j \text{ is active and} \\ & \text{module } k-1 \text{ is in resonance} \\ 0 & \text{otherwise} \end{cases}$$

  Whenever an input is presented, module $k-1$ is in resonance, and $|\mathbf{y1}_{k-1}| > \epsilon S$, then any $F2_k$ node j, such that $F2_{k-1}$ node h is active and $v_{k:jh} = 1$, is enabled and an initial category choice is made.

- *Category Choice*

  When module $k-1$ is in resonance, choose j such that $F2_k$ node j is enabled and

$$\frac{|\mathbf{w}_{k:j} \cap \mathbf{x}_k|}{\beta + |\mathbf{w}_{k:j}|} \tag{17}$$

  is maximal, with the result that $F2_k$ node j becomes active.

- *Category Match*

  A category matches an input when $F2_k$ node j is active, module $k-1$ is in resonance and

$$\frac{|\mathbf{w}_{k:j} \cap \mathbf{x}_k|}{|\mathbf{x}_k|} \geq \rho_k \tag{18}$$

  If the match is unsuccessful then $F2_k$ node j is disabled and another category choice is made.

  If the match is successful then learning takes place and the module is stable until the input is removed.

- *Learning*

  When $F2_k$ node j is active, and (18) holds then

$$\mathbf{w}_{k:j}^{new} = \eta(\mathbf{w}_{k:j}^{old} \cap \mathbf{x}_k) + (1-\eta)\mathbf{w}_{k:j}^{old}$$

  (adjustment of category prototype) and

$$\mathbf{v}_{k:j}^{new} = \mathbf{y2}_{k-1}$$

  whereby the association between a category and its chosen sub-category is created.

The HART-S training algorithm is shown in Figure 5. As a result of training, the network learns a hierarchical representation for the data set: each pattern in the set has a corresponding category – represented by active F2 nodes and links between them – at every layer in the hierarchy.

```
BEGIN HART-S
Initialise the network
DO
   FOR EACH instance (x ) in training set
      x_1 ← x ;; input to the first ART module
      k ← 1 ;; k is the index of the current ART module
      DO
         Search matching category (among prototypes w_{k:j}) for local input x_k
         IF existing matching category (J) found, i.e.,
               if w_{k:J} is "closest" to x_k and also |w_{k:J} ∩ x_k||x_k|^{-1} ≥ ρ THEN
            Learn local input in level k: w^{new}_{k:J} ← w^{old}_{k:J} ∩ x_k
         ELSE
            Create new category and w^{initial}_{k:new} ← x_k
         END IF
         Produce input to next layer: x_{k+1} ← x_k ∩ w^c_{k:J}
         k ← k + 1
      UNTIL k > L OR |x_k| < ε S ;; no more modules or local input too small
   END FOR
UNTIL Stability ;; i.e., when category prototypes do not change any further
END
```

Figure 5. The HART-S training algorithm as applied to a finite training set. The following parameters are used: L, the number of ART modules; $S = |\mathbf{x}|$, the (constant) norm of the binary input patterns; ρ, the single vigilance level (it can also be different in each module); ϵ, the resolution parameter, which controls the amount of sub-clustering in the network. The symbol ^{c} represents complement (or binary negation) operation.

3.2.2 Properties of HART-S

This section describes the main properties of the HART-S network. Here, we are making the further assumption that input patterns are normalised.

1. *Stability of Learning.*

   As with the HART-J network, all we need to do is show there is a lower bound on the size of local inputs for every ART module. There is such a bound since $\epsilon > 0$ and a module will only be activated when an input has a size larger than ϵS.

2. *Direct Access.*

   Once learning has stabilised direct access to category prototypes is assured for each ART module. As in HART-J, each module must wait for resonance in the previous module.

3. *Order of Presentation.*

   The order of presentation affects the clustering. Again this follows naturally from the properties of the individual ART modules.

4. *A Bound on the Number of Levels of Clustering.*

   The most levels of clustering possible is B when the following equation holds.

$$B = min\{k : \prod_{i=1}^{k}(1 - \rho_i) \leq \epsilon\} \tag{19}$$

Proof

Let the norm of the inputs be denoted S.

Let the size of the largest input possible to module k be I_k.

$I_1 = S$.

The smallest prototype formed in module k has size $\rho_k \; I_k$.

For $k > 1$

$$I_{k+1} = I_k(1 - \rho_k) \;=\; S\prod_{i=1}^{k}(1 - \rho_i).$$

The largest input to a possible module $k+1$ that will not activate it has size $S\epsilon$. So when $I_k \leq \epsilon$ module $k+1$ can never be activated.□

So $(1 - \rho_1) \leq \epsilon$ reduces the model to the standard ART model ($\epsilon = 1$ ensures this since $\rho_1 \geq 0$).

5. *Consistent Sub-clustering.*

   If an input pattern that is clustered at a level k of a clustering hierarchy is also clustered at level $k + 1$, then all inputs clustered with it at level k will also be clustered at level $k + 1$ if it exists.

The difference in size between a local input and its local prototype in a given module is the size of the input to the next module. So it is sufficient to show that the size of all local inputs clustered together in a given module is the same since then they will all generate the same sized input to the next module. It is the size of this input which determines if the next module will become active and sub-clustering occurs. All inputs clustered together in a given module must be clustered together in any previous modules, so are all of the same size since the inputs to module 1 are normalised.

6. *A Lower Bound on Effective Vigilance.*

$$\Omega_{k:min} = 1 - \prod_{i=1}^{k}(1 - \rho_i)$$

Proof

For any $\mathbf{X}$ and any clustering, if the left hand side of (16) is true so must be the right hand side. The worst case is when the size of $\mathbf{W}_k$ is as small as possible.

In this model, the size of $\mathbf{W}_k$ is the sum of the sizes of the local prototypes at level 1 to k, so clearly $\mathbf{W}_k$ is smallest when the size of the local prototypes in levels 1 to k is smallest.

The smallest prototype size at level 1 is $\rho_k|\mathbf{X}|$.

The smallest prototype size at level 2 is $\rho_2(1 - \rho_1)|\mathbf{X}|$.

The smallest prototype size at level k is $\rho_k \prod_{i=1}^{k-1}(1 - \rho_i)|\mathbf{X}|$.

So the smallest size for $\mathbf{W}_k$ is $(1 - \prod_{i=1}^{k}(1 - \rho_i))|\mathbf{X}|$.

So the greatest z such that the right hand side of (16) is true, is $(1 - \prod_{i=1}^{k}(1 - \rho_i))$. □

4 Experimental Results

We carried out three sets of experiments with the hierarchical ART networks presented here. The primary goal of each was:

1. To demonstrate the representational capabilities of HART-J and HART-S networks.
2. To explore the joint effect of vigilance (ρ) and resolution (ϵ) of the HART-S network on the depth of the resulting trees and on the training time.
3. To compare the classifications produced by HART networks to those of two classical hierarchical clustering methods and an AI conceptual clustering algorithm.

The details of the experiments and the results are discussed in the following sections.

4.1 Representational Capabilities of HART Networks

In order to demonstrate that the HART networks are capable of developing class *hierarchies*, we trained and tested the networks on the *Zoo* machine learning benchmark data set [9]. This data set contains 101 instances of animals described with 18 attributes such as "hair", "aquatic", "domestic" and so on. It is relatively small, but is adequate for the purposes of these experiments, which also allows us to make general observations about the behaviour of these models that will guide further analysis and development.

A number of simulations were run in a Unix/X windows environment using the neural network simulation program PlaNet [35] and programs written in C language. The binary attributes of the training samples were presented to the network in *complement coding* to assign significance to both the presence and absence of features, and to keep the input vectors normalised [14] to avoid proliferation of ART categories [37]. The only non-binary attribute ("number of legs") was presented in *generalised complement coding* [3]. The "type" attribute was ignored to eliminate any bias from the developed clusterings. The 18th attribute ("animal name") was simply used as a label for the individual instances. With this input encoding scheme, instances were presented to the network as 36-element binary vectors with the same norm of 16. In all the experiments, the HART networks were trained "off-line" on a training set containing all 101 instances of the "zoo" database. In each training session, the network was considered to be completely trained when each input pattern directly accessed a matching cluster at each active level of the cluster hierarchy (i.e., no category search or further learning occurred). Of particular interest are the effects of changes in vigilance and order presentation on clusterings and the comparative behaviour of the two models.

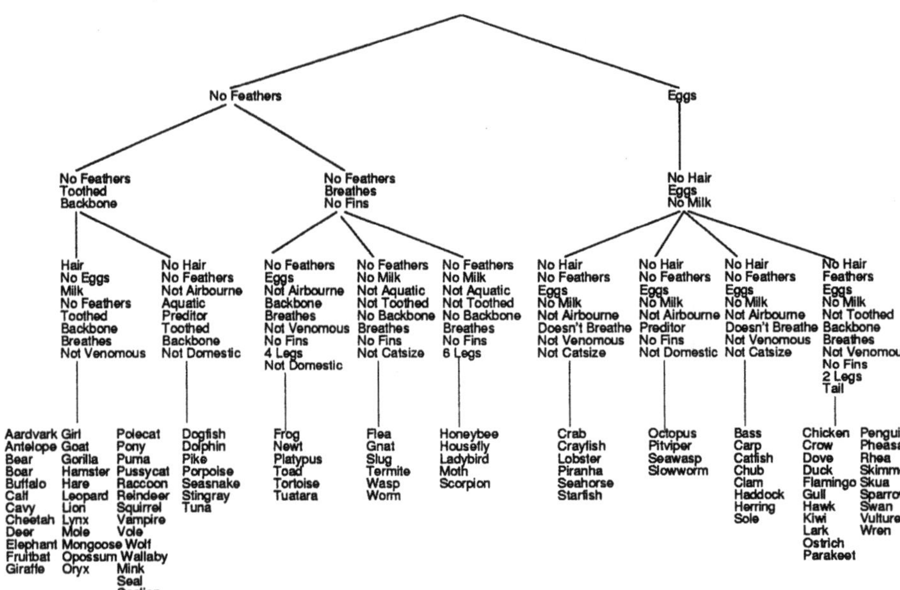

Figure 6. A cluster hierarchy learnt from the *Zoo* database by a three layer HART-J network. The vigilance levels were $\rho_1 = 0.45$, $\rho_2 = 0.3$ and $\rho_3 = 0.05$. Cluster prototypes specify all the features required for cluster membership and thus always increase in size as the hierarchy is descended. The root of the tree is only shown here for clarity, but is not actually represented in the network. In this example, ART modules 1, 2 and 3 developed 2, 3 and 9 categories, respectively. The tree structure "emerges" as input patterns are presented and the corresponding categories are activated at each level. The instances belonging to each leaf are shown here as the lowest layer (which could also have been learned by the network with an ART module inserted before the first one with a vigilance level of 1).

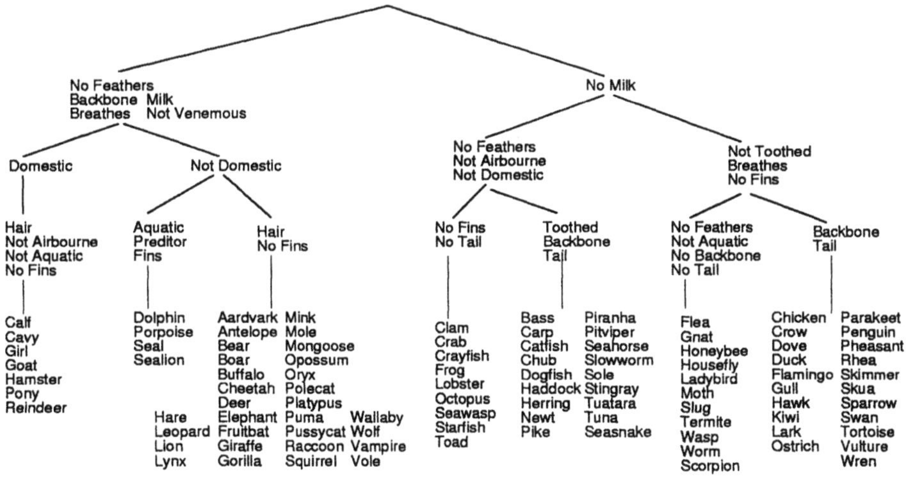

Figure 7. A cluster hierarchy learnt from the *Zoo* database by a three layer HART-S network. The resolution (ϵ) was 0.2 and all three vigilance levels were set to 0.06. Here the hierarchy is explicitly encoded in the network via the $\mathbf{v}_{k:j}$ associative links. Any particular feature is only used to specify a particular branch of the tree. The usage of the root and leaves is the same is in Figure 6.

Figures 6 and 7 are examples of hierarchies learnt by three-layer HART-J and HART-S networks, respectively. The choice parameters (β_k) and learning rates (η_k) were set to 1×10^{-5} and 1, respectively, in both cases. In general, we can observe that category prototypes are often duplicated in HART-J as prototypes are propagated on whether or not any clustering of them is achieved. Prototypes in HART-S by comparison always differ and furthermore information contained in one prototype is not copied in subcluster prototypes, as occurs in HART-J. A difficulty with HART-S is in predicting the number of F2 nodes needed in lower layers of a network due to the links created in hierarchical memory during learning. A higher setting of the resolution parameter tends to reduce the number of such nodes used to learn a particular set of inputs. Lastly we noted that, as with ART, the clusterings learnt by both HART models were affected significantly by a change in the order of input pattern presentation.

4.2 The Joint Effect of Vigilance and Resolution

This set of experiments explored the combined effect of vigilance (ρ) and resolution (ϵ) in HART-S on the depth of the resulting hierarchy and on required training time. The data set, training conditions and default parameter settings were identical to those used in the previous experiments.

Because of the significant effect of the order of input pattern presentation, each set of simulations consisted of 100 runs each with a different ordering of inputs and the measurements averaged out. The average deviations of these measurements indicate they should be quite reliable guides to the expected behaviour of HART-S. The depth of a hierarchy was taken as the deepest subcluster, and training time was taken as the number of presentations of the set of input patterns that caused any change of local category prototypes. Figures 8 and 9 show the average depth and training time (in epochs), respectively, while each of resolution and vigilance – which was the same for each module – was varied from 0 to 1 in steps of 0.1.

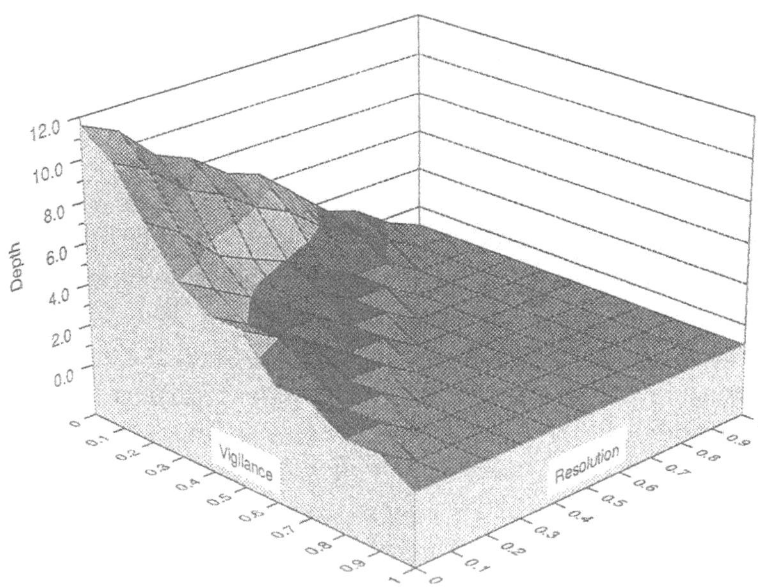

Figure 8. The average depth of HART-S hierarchies (vertical axis) with respect to vigilance and resolution (ρ and ϵ) after training with the *Zoo* database. Each point on the graph is the average of 100 clusterings with different orders of input pattern presentation. The flat area indicating only one level of clustering is formed where the network is constrained to behave as a single ART module. Lower settings of vigilance tend to result in smaller prototypes in any ART modules and thus more levels of clustering. Lower settings of resolution subcluster smaller differences with the same effect.

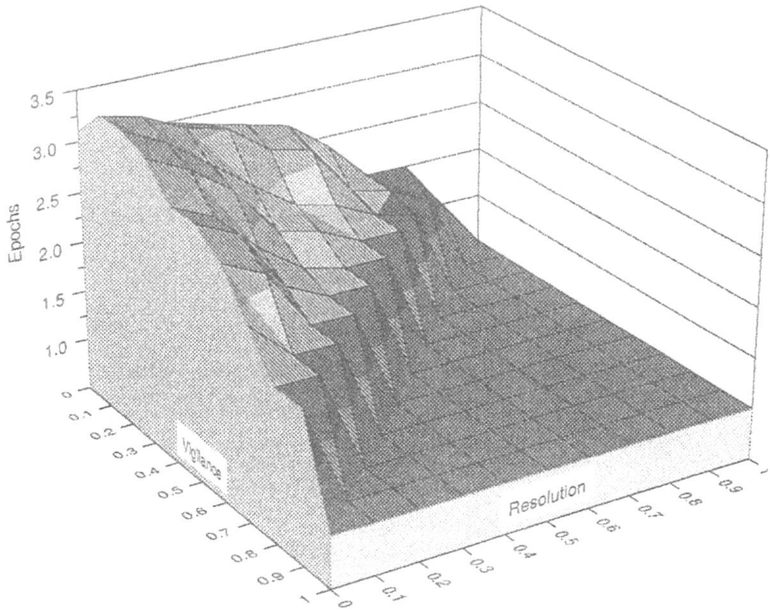

Figure 9. The average training time for HART-S hierarchies with respect to vigilance and resolution (ρ and ϵ) after training on the *Zoo* database. A unit on the vertical axis corresponds to one epoch in which the entire set of input patterns is presented. Each point on the graph is the average of 100 clusterings with different orders of input pattern presentation, these being the same ones as used in Figure 8.

The relationship expressed in equation (19) is clearly reflected in the first figure and in the second it is clear that the training time required to create a cluster hierarchy increases relatively slowly with increasing tree depth. Perhaps the symmetry of graphs suggest a redundant parameter. However, clusterings of the same depth and learnt in the same number of epochs but with significantly different combinations of ϵ and ρ are usually very different.

4.3 Comparison to Other Clusterings Methods

The primary motivation for designing these hierarchical ART networks was to have a modular network that is capable of fast and stable incremental learning of hierarchical clusterings of arbitrary sequences of input patterns by self-organisation. Having achieved this goal, we can now take a look at these networks from a different angle and view them simply as *implementations* of incremental clustering *algorithms*. The question that arises then is how these algorithms as such compare to other existing clustering methods. We attempted to find an answer to this with this set of experiments. We compared the clusterings gained by the HART-J and HART-S networks to standard statistical clustering methods and an incremental clustering machine learning method. The following two data sets from the UCI Repository [9] were used:

- *Soybean*: this data set contains 307 instances [5] described with 35 nominal attributes (and class information, which was ignored here). The attributes were encoded in generalised complement coding (see above) – with missing attribute values represented on a separate bit – yielding 134-element binary vectors of norm 35.
- *Mushroom*: this data set contains 1000 randomly selected examples from the original Mushroom set found in the repository. Each instance is described with 22 nominal features (and class information, which was ignored here). Again, attributes were encoded in generalised complement coding giving 126-element binary vectors (of norm 22) [6].

The two statistical hierarchical clustering methods that were used for comparisons are *compact* clustering (also known as *complete linkage*) and *group average* [22]. Both of them are agglomerative methods. In such methods, the initial configuration involves treating each instance as a cluster of size one. At any subsequent stage, two clusters are amalgamated into one using the two clusters judged closest by some suitable distance measure. Clustering ceases when all observations are combined into one cluster. In case of compact clustering, the distance between two clusters is taken as the maximum of the distances between any point of the first cluster and any point in the second cluster. Group average clustering is similar to compact clustering, differing only in the criterion by which extant clusters are to be fused at any given stage. Here the criterion is the average of the distances between points in the two clusters. The

[5] There are two versions of this data set in the UCI repository: *soybean-small* and *soybean-large*. We used the latter.

[6] This is exactly the same encoding that was used in [14].

built-in library functions of the S-PLUS statistical package were used here.

While the above clustering methods are non-incremental (i.e., the whole data set has to be available for the computation), several incremental machine learning algorithms have been proposed for unsupervised concept formation. One such method is called *Froggit*[7]. Froggit is an incremental, top-down, hierarchical clustering algorithm derived from COBWEB [19] and CLASSIT [21]. Like COBWEB, it processes new instances, one at a time, incorporating them into a class hierarchy, and attempting to incrementally optimise the hierarchy. Froggit differs from COBWEB and CLASSIT in three significant ways. First, Froggit uses an information theoretic measure of the goodness of a partition in its optimisation search. This measure allows Froggit to handle both numeric and discrete attributes in a well-founded way. The measure is based on a minimum message length approach [43], but is not completely rigorous. Second, the search strategy in the optimisation phase is much richer than COBWEB's. It uses a beam hill-climbing search, and a wider range of hierarchy modification operators than COBWEB. These allow it to smooth the search space and avoid some of the local minima that trap COBWEB. Third, Froggit separates the optimisation from the incorporation, which not only improves the effectiveness of the search, but also allows a wider range of incorporation strategies [2].

Since the class representation of all of the above methods (including HART) are markedly different, a reasonable way to compare the methods is to look at their partitionings at the instance level (instead of comparing their class descriptions). The developed clusterings of the HART networks, however, are affected by the order of input presentation. Therefore, instead of determining a single similarity value, we can estimate the probability with which categories of HART networks and other methods agree. More precisely, given the same number of categories found, we estimate

$$P_{hd}(c_{ij} = 1 | h_{ij} = 1), \quad i < j, \ i, j = 1, \ldots D$$

[7] An earlier version of Froggit, called *DySect*, has been applied successfully to data analysis [18].

Table 1. This table shows estimates of "agreement" probabilities between HART-J and three other clustering methods on the Soybean and Mushroom data sets (P_{hs} and P_{hm} columns, respectively, for each method). The *'s indicate that Froggit could not find exactly N_c categories, in which case the probabilities were taken from a partition having the closest number of categories.

		Compact		Group		Froggit	
N_c	P_r	P_{hs}	P_{hm}	P_{hs}	P_{hm}	P_{hs}	P_{hm}
5	0.200	0.207	0.313	0.898	0.926	0.498*	0.346
10	0.100	0.112	0.236	0.532	0.794	0.474	0.768*
15	0.067	0.094	0.231	0.291	0.703	0.540	0.887
20	0.050	0.067	0.105	0.167	0.657	0.376*	0.501*

where D is the size of the data set d, $c_{ij} = 1$ and $h_{ij} = 1$ if instances i and j are assigned to the same class by a clustering algorithm and a HART network in a given training run, respectively. In other words, P_{hd} is the probability that if instances i and j are found to be in the same category by a HART network, they are also found to be in the same category by a clustering algorithm. The best case is when $P_{hd} = 1$, i.e., the two methods agree 100% of the time. The worst case is complete randomness, or non-correlation between the methods. However, even in this case, there is a finite probability that the random selection of categories will agree with any clustering method. In case of complete random categories, this probability (P_r) is $\frac{1}{N_c}$ where N_c is the number of classes in the given partition. So clearly, $P_r \leq P_{hd} \leq 1$.

Table 1 and 2 show the estimated P_{hd} probabilities for the HART-J and HART-S networks, respectively. The number of classes (N_c) were gained differently in each method. For the statistical clustering methods (Compact and Group), a cut of the tree was taken at the desired level. For Froggit, one of its parameters, which controls the specificity (and therefore the number) of top-level classes in the hierarchy, was varied until the desired number was produced. In HART-J, different number of classes were gained by simply observing the partitionings of different modules in the ART cascade. In HART-S, the appropriate cut of the tree was made through the use of the resolution parameter (ϵ) in a trained network (varying it between 0 and its value at training), which typically resulted in cuts spanning multiple ART modules.

Table 2. This table shows estimates of "agreement" probabilities between HART-S and three other clustering methods on the Soybean and Mushroom data sets. For more explanation, refer to Table 1.

		Compact		Group		Froggit	
N_c	P_r	P_{hs}	P_{hm}	P_{hs}	P_{hm}	P_{hs}	P_{hm}
5	0.200	0.206	0.313	0.897	0.925	0.532*	0.470
10	0.100	0.111	0.237	0.534	0.791	0.430	0.688*
15	0.067	0.093	0.232	0.304	0.700	0.485	0.751
20	0.050	0.069	0.105	0.173	0.659	0.341*	0.438*

Table 3. The comparison of three ART-based methods and Froggit against the three "base" ones used above. The numbers are the same P_{hd} probabilities used in Tables 1 and 2. Here we show results for 15 classes (i.e., $P_r = 0.067$) learned from the Mushroom database.

	Compact	Group	Froggit
ART	0.231	0.698	0.724
HART-J	0.231	0.703	0.887
HART-S	0.232	0.700	0.751
Froggit	0.232	0.705	1.000

First, the results show that the HART networks produce clusterings that are correlated (at least according to our measure) to the other methods at better (sometimes much better!) than chance level. It can also be seen clearly that the numbers for HART-J and HART-S are very similar suggesting that even though the architectures and developed class representations are different, they produce similar clusterings. The similarity between HART and the other is noticeably higher for the Mushroom data set. The Group method and Froggit are apparently more similar to HART than the Compact method. For the two classical clustering methods, the level of agreement tends to increase as N_c decreases.

We also ran some experiments to gauge the similarity of clusterings of single ART networks to those of HART-J, HART-S and Froggit (against the three "base" methods used in the tables above) as well as to see how Froggit compares to the two statistical clustering methods. Table 3 shows the results on the Mushroom database when the number of classes was 15.

A very interesting result is that the HART and ART networks (and Froggit) show remarkably similar levels of agreement with the compact and group average methods. This seems to suggest that the hierarchical ART networks offer their added representational power without any apparent change in their clustering behaviour. Also, the HART networks and Froggit appear to be closer to each other than to the classical methods while they are also slightly closer to Froggit than the ART network. Recall, however, that the aim has been to provide a modular neural network model that inherits the clustering properties of some kind of ART type components and integrates them to perform the task of discovering a set of useful partitions and a hierarchical relationship between them that is both consistent and usable (rather than to find a hierarchical clustering method better than, for example, Froggit).

5 Optimisation of HART-S Clustering

The known drawbacks of ART networks are (i) category proliferation [37], (ii) their inability to represent (and learn) structure (hierarchical or other) that might be present in the environment, and (iii) their sensitivity to the order in which inputs are presented from a training set. The latter often manifests itself in the network creating many unnecessarily complex categories/rules and failing to discover statistically significant relationships.

A commonly used solution to category proliferation is to present inputs in complement coding [15]. Hierarchical ART architectures have been proposed to extend the knowledge representation capabilities of single ART networks [28], [5], [6]. Very little work has been done, however, to overcome the third drawback.

This section addresses the problem of category (or rule) complexity in ART networks, and describes our attempts to alleviate it by introducing optimisation methods into the HART-S hierarchical ART network (introduced in Section 3.2). The optimisation steps are intertwined with the learning process, whereby some additional computational cost is traded off for an improved (more robust) set of categories (or rules) while maintaining the main properties of HART-S like fast, stable and incremental learning of category hierarchies.

5.1 Two Optimisation Methods for HART-S

The following is a description of the two optimisation methods we devised.

Cluster size threshold (CST) The development of this method was motivated by the observation that HART-S (and a single ART, too, for that matter) often creates many small categories initially, many of which which are either abandoned later or remain small and yet do not represent an important exception in the training set. When this method is applied, clusters below a certain size are deleted from the network (and the corresponding nodes are freed up for re-use). This also means that if a non-terminal (or leaf) cluster is deleted, all of its children are also deleted from the tree. The method has a control parameter, which is the size of the cluster as a percentage of all the instances in the training set. This value is high initially, and is gradually decreased to zero (which corresponds to no deletion of categories).

Prototype rebuilding (PRB) This method is based on the observation that, as a result of the self-stabilisation process of HART-S (and ART) networks, category prototypes can contain feature values that none of the instances in the cluster they represent have. The network, due to its unidirectional weight adjustment (which is important for stability), can never restore these values by itself. This is undesirable since prototypes may not be entirely representative of their cluster at the end of training (which is especially important if the goal is to obtain descriptive rules of input clusters). When this method is applied, if a (binary) feature value is found to be 0 in the prototype whereas all instances in that cluster have 1's, that feature is "re-built" by setting the corresponding feature value to 1. The method has a parameter that controls the amount of rebuilding by specifying the minimum number of features needed to be rebuilt in a prototype for it to be actually changed (re-built). This parameter is zero initially, and is gradually increased to the length of the prototype vector (which corresponds to no rebuilding of categories).

The application of the above methods (either separately or combined) is done in "optimisation steps" carried out periodically during learning. We have used two simple schedules: one in which the HART-S network was trained completely (i.e., until stability was reached), after which a single optimisation step was applied, and then the network was trained until convergence again, and so on; in the other, the optimisation step was applied after every training epoch (i.e., presentation of the entire training set) during training until stability was reached. We call the two versions "s" (for stability) and "c" (for cyclic), respectively.

5.2 Experimental Results

In order to test the effectiveness of the above methods, we carried out experiments using the *soybean* and *mushroom* data sets that were described in Section 4.

We measured the effect of the optimisation methods by using the following set of metrics:

- *Epochs*: the total number of times the entire data set is presented until training converges.
- *Nodes*: the total number of category nodes developed by the network (i.e., total number of categories in the hierarchy).
- *Leaves*: the total number of leaves (or terminal nodes) in the developed tree (or category hierarchy).
- *Branching factor*: the average branching factor of the tree.
- *Prototype size*: the average norm (i.e., number of active or critical features) of all prototypes in the network.
- *Missed features*: the average number of prototypical features in the tree that could be active, but are not (see the PRB method above).
- *Prototypical variance*: a measure of the variability of prototypical features as a number between 0 (always the same prototypes are found) and 100 (prototypes are found at random).

For each method and metric used, 100 training runs were carried out, each with a different random order of presentations of all the instances

Table 4. This table shows the effect of each optimisation method applied independently. The HART-S method is the standard one without optimisation. The numbers are averages over 100 training runs. The vigilance (ρ) and resolution (ϵ) parameters were fixed at 0.1 and 0.5, respectively. Columns "S" and "M" for each metric represent the Soybean and Mushroom-1000 databases, respectively.

	Number of Epochs		Number of Nodes		Number of Leaves		Branching factor	
Method	S	M	S	M	S	M	S	M
HART-S	2.98	1.89	83.9	62.3	35.8	31.2	2.07	2.16
CST-s	21.70	16.20	58.9	46.2	32.8	26.5	2.18	2.23
CST-c	7.62	6.83	61.3	47.0	34.1	27.0	2.18	2.24
PRB-s	16.40	11.40	57.1	41.4	28.0	23.1	2.22	2.24
PRB-c	5.83	5.09	63.0	45.6	29.3	24.6	2.16	2.16

	Prototype size		Missed features		Prototypical variance	
Method	S	M	S	M	S	M
HART-S	34.1	33.8	26.6	26.0	9.11	6.04
CST-s	34.2	34.6	24.1	24.1	8.67	5.25
CST-c	34.5	34.6	24.2	24.3	9.00	5.32
PRB-s	35.7	36.1	19.8	20.6	9.13	7.57
PRB-c	34.4	34.4	21.9	22.7	8.98	7.40

in the data set. In each training session, the network was trained until no changes to the category prototypes were detected for an entire epoch.

The results are shown in Table 4.

We observe that each method results in about 30% less number of category nodes on both databases. What is perhaps a little unexpected is that the PRB method scores at least as well as (and sometimes better than) the CST method on the number of nodes. The branching factor is increased slightly (while the depth of the tree remained at about 6) resulting in more compact trees in general. As expected, the PRB method scores better on the "missed features" metric, but also tends to increase the "prototypical variance" slightly (which CST consistently reduces). It is also evident that the "cyclic" schedule (or "-c" versions) provides a better trade-off since it only performs slightly worse on the metrics used here at a significantly less extra computational cost (as expressed in the number of epochs).

6 Discussion

In the following, we discuss other neural network models that we consider to be more closely related to the hierarchical ART networks introduced here. We shall focus on identifying similarities and differences.

Lampinen and Oja proposed a model, in which a hierarchy of self-organising maps (SOMs, see [31]) was used for visual feature extraction [32]. In a MSOM network, there is a separate input vector for each SOM network in the first layer. The input to the second layer is a vector with the *index* of the winning node in each SOM network of the first layer. The authors admit this does not bear any resemblance to real networks, but provides a useful and effective method for feature extraction in their application. This hierarchy is rather different from that of a HART network, both in what input each layer receives and how the layers are interconnected. In particular, the hierarchy in a HART is not encoded explicitly in the network architecture, which emerges (via learned links) as the network undergoes training.

Li et al. proposed a "neural tree" architecture for hierarchical classification and vector quantisation [34]. It implements a splitting (or divisive) mechanism by self-organisation, like HART-S. However, in a neural tree, each node in every layer receives the same (global) input pattern, and the layers and nodes are selected during learning and recall using a global control algorithm. In HART, only the lowest layer (or module 1) receives the original input pattern, and its control is completely autonomous, and emerges as a result of interaction between layers and nodes using only local operations. After training, the neural tree is optimised by pruning and splitting taking ideas from machine learning (e.g. CART [10]). The neural tree was applied to data compression and speech processing.

Despite their usefulness in practical applications, the above networks would not be suitable in autonomous learning tasks because of slow learning and instability against changes in the environment (see for example [37]). Partly motivated by these problems, researchers have looked at hierarchical ART systems in recent years.

The ART3 architecture [13] can be considered as the first example of hierarchical ART networks. Carpenter et al. define two modular ART-based models ("ART bidirectional hierarchy" and "ART cascade") for learning hierarchical clusterings, in which there is a close coupling between each ART module mediated by special gate control signals, and each module learns information that is already processed by previous layers. This mechanism is very similar to HART except for one main difference: in ART3, each component network performs *contrast enhancement* [12] in the F2 layer instead of selecting a winner. The output pattern of the *F2 layer* is then used as input to the next ART layer. In contrast, information that is passed on to successive layers in a HART network is based on *prototypes* (either directly or as a difference vector) rather than *distances* of inputs from the category prototypes (like in ART3). Also, because in ART3 each F2 layer is set to produce partially compressed codes via nonlinear differential equations, calculating the output pattern is computationally more expensive than in HART, where categories are accessed directly in a trained network resulting in faster recall of learned prototypes.

An interesting and seemingly closely related ART-based hierarchical model is the "ART tree" [17] that was designed to be embedded in a Neural Information Retrieval System (NIRS) to facilitate the intelligent and fast retrieval of engineering design. The ART tree is a collection of "ART1 macrocircuits" connected in a tree topology. Layers in the tree are trained successively by first breaking up the set of designs into a few general categories, and then training the modules in lower layers on only those patterns that belong to their respective parent class. The main differences between the ART tree and HART are that in an ART tree (i) each macrocircuit is allowed to "see" the entire (global) input pattern, (ii) the hierarchy is explicitly encoded in the network prior to learning in the form of a tree and (iii) learning and recall are under strict global control as part of the NIRS CAD system. Also, the ART tree – however generic in principle – is application specific (with different layers being selective to given features only), whereas HART does not make any assumptions about the relative importance of input features.

Soon and Tan proposed the concept hierarchy network (CHN, in [41]) as an ART-based memory model that forms concept hierarchy by learning sample relations between concepts. All concepts are represented in a concept layer. Relations between a concept and its defining lower level concepts are chunked as cognitive codes represented in a coding layer. This network is a good example of how ART networks can be combined to learn more complex tasks, which individual components cannot do by themselves. However, even though the network is modular and learns a hierarchy, these cannot be seen as the main design principle in either the architecture or the learning mechanism. It shows up, for example, in how the CHN is limited to two-level hierarchies whereas in HART the number of levels is variable and depends on the training set and the vigilance level of component ART modules.

The first of our ART-based models (SMART, for "Self-consistent Modular ART") was introduced in [3], in which layers in a hierarchy were connected through an associative layer much the same way as the two ART modules are connected in an ARTMAP network [14]. In the SMART network, each module learned to cluster the original inputs at different levels of vigilance, and self-consistency was maintained via ARTMAP's "match tracking" process. This is a rather different alternative approach to ART-based hierarchical clustering resulting in markedly different developed hierarchies [4].

More recently, Ishihara et al. introduced "arboART" [28], an ART-based architecture, which is identical in basic principles to our earlier model (HART-J). In other words, both arboART and HART-J develop class hierarchies by clustering the prototypes developed at the lower level. The only significant difference is that arboART is based on the "ART1.5-SSS" network [27], which is an adaptation of ART1 to analogue patterns for small sample size. The clustering performance of arboART was found to be comparable to that of a the centroid method in statistical clustering. The existence of arboART and HART-J demonstrates well the advantage of such architecture, where different types of ART networks can be "plugged in" to achieve certain goals according to different criteria.

We also note that the HART networks possess certain features of self-similar fractal-like architectures, which are also advantages of modular networks [23]. Namely, ART modules in a HART-J or HART-S network interact via modifiable links as well as fixed (non-specific) control signals, similar to the way layers and individual nodes interact within a single ART network. This self-similarity can be carried even further in the

HART-S network, in which training starts with a pool of uncommitted ART modules that are later allocated as needed depending on the training set and a global parameter ("resolution", in Section 3), much like individual nodes become committed in a single ART module depending on a within-ART global parameter (vigilance). This opens up a number of possibilities for creating more sophisticated interactions between ART layers to direct its search in the hierarchy and enhance its performance. (Another good example for combining modularity and hierarchy in a self-similar structure is [29].)

As a final note, the HART-S network essentially implements the same kind of hierarchical search and recall that was discussed in [1]. However, the recall there was achieved by temporal sampling of the same input signal (dog's sniffing) until the network found the most specific category that described the input. Because of the reliance of repetitive sampling, that network was able to represent the hierarchy in a single layer (with appropriate settings of network dynamics). So we can view the HART-S network as having explicit spatial representation of the hierarchy, which "compensates" for the lack of repetitive sampling.

7 Conclusions

This chapter described a modular neural network architecture (HART-S) that is capable of stable and fast learning of hierarchical clusterings of input sequences by dividing initially large classes into smaller, more specific ones. The network is a cascade of Adaptive Resonance Theory modules, each clustering the *differences* between the input to the previous module and its selected category prototype. The class hierarchy is learned via associative links between F2 layers of neighbouring modules. The depth of the hierarchy can be controlled with a global parameter called resolution. We have shown that the network, like the "joining" Hierarchical ART we also discussed here, retains the useful properties of the component ART networks. The experiments demonstrated the representational and learning capabilities of both the HART-J and HART-S networks, and also suggested that the developed clusterings might be closely related those produced by some classical and/or machine learning clustering algorithms. We believe investigating such similarities (and differences too) further could provide us with more insights into the advantages and drawbacks of using ART-based modular networks in real-world problems. This may also allow us to address issues such as extracting rules from trained HART networks, and/or optimising the developed hierarchies. Along the latter line, it may be useful to view HART networks (especially in fast learning mode) as implementing the first "sorting" stage in a multi-stage clustering mechanism [20].

We also note that as a consequence of the modular architecture of the HART-S and HART-J networks, it is relatively simple to modify them to achieve slightly different functionalities. One example for this is the FUZZY HART-S network [7] in which the ART1 components are replaced with Fuzzy ART [15] modules. This way, the network can learn from input patterns represented on binary as well as fuzzy attributes [44].

The HART-S network with the added optimisation steps can also be likened to machine learning algorithms where the clusterings are found via the iterative application of a quick "sorting" and a successive "optimisation" stage [20]. We also suggest that the proposed optimisation methods, besides being useful for improving the extracted categories and rules from finite data sets, can be implemented as a background (network) process in autonomous learning systems allowing them to learn from the environment quickly and then continuously optimising the data representation for effective storage and utilisation. It is also worthwhile investigating further the combination of these and possibly other methods as well as the issue of selecting an optimal schedule for the optimisation to achieve the best trade-off.

Finally, we believe the hierarchical networks introduced here are not only good examples of modularity, but can also suggest possible mechanisms for learning complex recognition tasks in real and artificial neural networks. We also believe this model is worth investigating further, which could lead us to a refined model capturing some of the more complex phenomena that occur in both artificial and biological systems.

Acknowledgments

This chapter is based largely on [6], which is reproduced here with the permission of Carfax Publishing Limited, P.O. Box 25, Abingdon, Oxfordshire, OX14 3UE, United Kingdom. Other portions (in Section 5, which appeared in [8]) are reprinted, with permission, from Proceedings of IEEE International Joint Conference on Neural Networks (IJCNN'98), pages 2352–2356; ©1998 IEEE.

This work was in part supported by Victoria University of Wellington (Research Grants No. RGNT/596/303 and No. RSCH/596/582) as well as the Computer Science Group at VUW. We wish to acknowledge the programming support of David Gilligan. We also thank the reviewers for their comments, which helped us improve the papers.

References

[1] Ambros-Ingerson, J., Granger, R., and Lynch, G. (1990), "Simulation of paleocortex performs hierarchical clustering," *Science*, vol. 274, pp. 1344-1348.

[2] Andreae, P. (1996), "Froggit's goodness measure and search strategy," Personal communication.

[3] Bartfai, G. (1994), "Hierarchical clustering with ART neural networks," In *Proceedings of the IEEE International Conference on Neural Networks*, IEEE Press, vol. 2, pp. 940-944.

[4] Bartfai, G. (1995), "A comparison of two ART-based neural networks for hierarchical clustering," In N. Kasabov and G. Coghill, editors, *ANNES'95, Proceedings of the Second New Zealand International Two-Stream Conference on Artificial Neural Networks and Expert Systems*, IEEE Computer Society Press, pp. 83-86.

[5] Bartfai, G. (1996), "An ART-based modular architecture for learning hierarchical clusterings," *Neurocomputing*, vol. 13, no.1, pp. 31-46.

[6] Bartfai, G. and White, R. (1997), "Adaptive resonance theory-based modular networks for incremental learning of hierarchical clusterings," *Connection Science*, vol. 9, no. 1, pp. 87-112.

[7] Bartfai, G. and White, R. (1997), "A Fuzzy ART-based modular neuro-fuzzy architecture for learning hierarchical clusterings," In *Proceedings of the IEEE International Conference on Fuzzy System*, pp. 1713-1718.

[8] Bartfai, G. and White, R. (1998), "Learning and optimisation of hierarchical clusterings with ART-based modular networks," In *Proceedings of IEEE International Joint Conference on Neural Networks (IJCNN'98)*, pp. 2352-2356.

[9] Blake, C., Keogh, E., and Merz, C.J. (1998), "UCI repository of machine learning databases," http://www.ics.uci.edu/~mlearn/MLRepository.html.

[10] Breiman, L., Friedman, J.H., Olshen, R.A., and Sonte, C.J. (1984), *Classification and Regression Trees*, Wadsworth and Brooks, Monterey, CA.

[11] Carpenter, G.A. and Grossberg, S. (1987), "ART2: Self-organizing of stable category recognition codes for analog input patterns," *Applied Optics*, vol. 26, no. 23, pp. 4919-4930.

[12] Carpenter, G.A. and Grossberg, S. (1987), "A massively parallel architecture for a self-organizing neural pattern recognition machine," *Computer Vision, Graphics, and Image Processing*, vol. 37, pp 54-115.

[13] Carpenter, G.A. and Grossberg, S. (1990), "ART3: Hierarchical search using chemical transmitters in self-organizing pattern recognition architectures," *Neural Networks*, vol. 3, pp. 129-152.

[14] Carpenter, G.A., Grossberg, S., and Reynolds, J.H. (1991), "ARTMAP: Supervised real-time learning and classification of nonstationary data by a self-organizing neural network," *Neural Networks*, vol. 4, pp. 565-588.

[15] Carpenter, G.A., Grossberg, S., and Rosen, D.B. (1991), "Fuzzy ART: Fast stable learning and categorization of analog patterns by an adaptive resonance system," *Neural Networks*, vol. 4, pp. 759-771.

[16] Carpenter, G.A. and Grossberg, S. (1988), "The ART of adaptive pattern recognition by a self-organizing neural network," *IEEE Computer*, vol. 21, no. 3, pp. 77-88.

[17] Caudell, T.P., Smith, S.D.G., Escobedo, R., and Anderson, M. (1994), "NIRS: Large scale ART-1 neural architectures for engineering design retrieval," *Neural Networks*, vol. 7, no. 9, pp. 1339-1350.

[18] Dawkins, B.P., Andreae, P.M., and O'Connor, P.M. (1994), "Analysis of olympic heptathlon data," *Journal of the American Statistical Association*, vol. 89, no. 427, pp. 1100-1106.

[19] Fisher, D.H. (1987), "Knowledge acquisition via incremental conceptual clustering," *Machine Learning*, vol. 2, no. 2, pp. 139-172.

[20] Fisher, D. (1996), "Iterative optimization and simplification of hierarchical clusterings," *Journal of Artificial Intelligence Research*, vol. 4, pp. 147-179.

[21] Gennari, J.H., Langley, P., and Fisher, D. (1989), "Models of incremental concept formation," *Artificial Intelligence*, vol. 40, pp. 11-61.

[22] Gordon, A.D. (1981), *Classification: Methods for the Exploratory Analysis of Multivariate Data*. London: Chapman and Hall.

[23] Happel, B. and Murre, J. (1994), "Design and evolution of modular neural network architectures," *Neural Networks*, vol. 7, no. 6/7, pp. 985-1004.

[24] Hartigan, J.A. (1975), *Clustering Algorithms*. John Wiley & Sons, Inc.

[25] Hecht-Nielsen, R. (1987), "Counterpropagation networks," *Applied Optics*, vol. 26, no. 23, pp. 4979-4984.

[26] Hrycej, T. (1992), *Modular Learning in Neural Networks; A Modularized Approach to Neural Network Classification*, Sixth-Generation Computer Technology Series, John Wiley & Sons, Inc.

[27] Ishihara, S., Ishihara, K., Nagamachi, M., and Matsubara, Y. (1993), "ART 1.5-SSS for kansei engineering expert system," In *Proceedings of International Joint Conference on Neural Networks*, pp. 2512-2515.

[28] Ishihara, S., Ishihara, K., Nagamachi, M., and Matsubara, Y. (1995), "arboART: ART based hierarchical clustering and its application to questionnaire data analysis," In *Proceedings of the IEEE International Conference on Neural Networks*, IEEE Press, vol. 1, pp. 532-537.

[29] Jordan, M.I. and Jacobs, R.A. (1992), "Hierarchies of adaptive experts," In J. E. Moody, S. J. Hanson, and R. P. Lippmann, editors, *Advances in Neural Information Processing Systems 4*, Morgan Kaufmann, San Mateo, CA, pp. 985-992.

[30] Jordan, M.I. and Jacobs, R.A. (1993), "Hierarchical mixtures of experts and the EM algorithm," Technical Report 9203, MIT Computational Cognitive Science, MIT, Cambridge, MA.

[31] Kohonen, T. (1982), "Self-organized formation of topologically correct feature maps," *Biological Cybernetics*, vol. 43, pp. 59-69.

[32] Lampinen, J. and Oja, E. (1992), "Clustering properties of hierarchical self-organizing maps," *Journal of Mathematical Imaging and Vision*, vol. 2, pp. 261-272.

[33] LeCun, Y., Boser, B., Denker, J.S., Henderson, D., Howard, R.E., Hubbard, W., and Jackel, L.D. (1990), "Back-propagation applied to handwritten zip code recognition," *Neural Computation*, vol. 1, pp. 541-551.

[34] Li, T., Fang, L., and Li, K.Q-Q. (1993), "Hierarchical classification and vector quantization with neural trees," *Neurocomputing*, vol. 5, pp. 119-139.

[35] Miyata, Y. (1991), *PlaNet, A Tool for Constructing, Running, and Looking into a PDP Network.*

[36] Moody J. and Darken, C.J. (1989), "Fast learning in networks of locally-tuned processing units," *Neural Computation*, vol. 1, pp. 281-294.

[37] Moore, B. (1989), "ART1 and pattern clustering," In D. Touretzky, G. Hinton, and T. Sejnowski, editors, *Proceedings of the 1988 Connectionist Models Summer School*, Morgan Kaufmann, San Mateo, CA, pp. 174-185.

[38] Oja, E. (1989), "Neural networks, principal components, and subspaces," *International Journal of Neural Systems*, vol. 1, pp. 61-68.

[39] Rumelhart, D.E., McClelland, J.L., and the PDP Research Group (1986), *Parallel Distributed Processing; Explorations in the Microstructure of Cognition*, the MIT Press, vol. 1: Foundations, chapter 5, pp. 151-193.

[40] Shepherd, G.M. (1974), *The Synaptic Organization of the Brain.* Oxford University Press, New York.

[41] Soon, H. and Tan, A. (1993), "Concept hierarchy network for inheritance systems: Concept formation, property inheritance and conflict resolution," In *Proceedings of 15th Annual Conference of Cognitive Science Society*, Lawrence Erlbaum Associates, pp. 941-946.

[42] Van Essen, D.C., Anderson, C.H., and Felleman, D.J. (1992), "Information processing in the primate visual system: An integrated systems perspective," *Science*, vol. 255, no. 5043, pp. 419-423.

[43] Wallace, C.S. and Boulton, D.M. (1968), "An information measure for classification," *The Computer Journal*, vol. 11, no. 2, pp. 185.

[44] Zadeh, L.A. (1965), "Fuzzy sets," *Information and Control*, vol. 8, pp. 338-353.

CHAPTER 6

ART-BASED AUTONOMOUS LEARNING SYSTEMS: PART I – ARCHITECTURES AND ALGORITHMS

C.P. Lim
School of Industrial Technology
Universiti Sains Malaysia
11800 Penang
Malaysia

R.F. Harrison
Department of Automatic Control and Systems Engineering
The University of Sheffield
Mappin Street, Sheffield S1 3JD
United Kingdom

This chapter describes the design of novel ART-based intelligent systems that are able to learn and, at the same time, to refine their knowledge in perpetuity. Fuzzy ARTMAP and the Probabilistic Neural Network are integrated to form a hybrid system that possesses the desirable properties for incremental, causal learning as well as for Bayesian probability estimation. Subsequently, a multiple neural network architecture is devised to aggregate outputs from several individual networks into a unified decision. A number of algorithms is proposed to increase the generalization and adaptability of the resulting systems.

1 Introduction

Over the past few years, artificial neural networks have been used as an "intelligent" problem-solving tool in various disciplines of science and technology, and promising results have been reported for a wide range of applications. Although many characteristics of neural networks have been studied, one aspect that receives little attention, and yet may be important to certain intelligent systems, is the ability to absorb information continually and autonomously without corrupting or forgetting previously acquired information. This ability is desirable especially for systems operating in changing environments so as to perform their roles with a higher degree of autonomy and with less human intervention. Therefore, we focus our work on the development of neural-network-based intelligent systems that are able to learn autonomously and, at the same time, to refine their knowledge perpetually in response to an arbitrary sequence of input patterns in non-stationary (time-varying) as well as stationary (time-invariant) environments.

Perhaps the most commonly explored neural network architecture is the feedforward type of neural networks such as the Multi-Layer Perceptron (MLP) [1], and the Radial Basis Function (RBF) network [2]. Although many useful properties of feedforward networks have been proved, *e.g.*, they are universal function approximators [3], [4] and their outputs are estimates of the Bayesian (*a posteriori*) probabilities [5], [6], the "standard" learning procedure of these networks is essentially an *off-line* process. It starts off with a training phase in which information on the environment (provided by a set of input-output pattern pairs) is encoded into the network by adapting the connecting weights using some learning algorithm. Then, performance of the network is evaluated with a validation data set. If a satisfactory performance is attained, the network is put into operation, and no further weight adaptation (or learning) takes place. This approach is viable only when the environment is stationary and the training data is sufficiently representative. Nevertheless, if the environment changes and the network is presented with new, previously unseen data, there is generally no built-in mechanism for it to recognize these novel patterns. In order to accommodate current information about the environment, the network has to be re-trained using a mixture of new and old data. This drawback, suffered by most neural network models, arises from the so-called *stability-plasticity* dilemma [7]. The fundamental questions posed by the dilemma are (i) how can a learning system remain plastic, or adaptive, in response to significant events, and yet remain stable in response to irrelevant events? (ii) how can a system adapt to new information without corrupting or forgetting previously learned information? [8], [9]

To tackle the stability-plasticity dilemma, the *Adaptive Resonance Theory* (ART) family of networks [8]-[15] has been developed. An ART network is an incremental learning system that self-organizes and self-stabilizes in response to an arbitrary sequence of patterns in non-stationary environments. It applies a sequential clustering algorithm to establish a set of prototype patterns for input vectors that are clustered closely together in the input space. Each ART network also includes a novelty detector that measures against a threshold the similarity between the prototype patterns stored in the network and the current input pattern. When the similarity criterion is not satisfied, a new node can be created in the network with the input pattern coded as its prototype pattern. As a result, the number of nodes grows with time, subject to the similarity criterion, in the process of learning a good network configuration incrementally. As different tasks demand different capabilities from the network, this dynamic network architecture and autonomous learning methodology thus avoids the need to pre-specify a (static) network size or to re-train the network in non-stationary environments.

The success of ART as an incremental learning system may be ascribed to the following: it is stable against noisy inputs, attentive to novel patterns and, at the same time, resistive to memory erosion [8]. Additionally, there are few arbitrary parameters to be set by the user. ART's self-organizing characteristics offer a high degree of autonomy and enable systems to learn continually *in situ* with minimum intervention after the design stage. As a result, ART opens up the way for developing autonomous learning systems that may ultimately allow domain experts, rather than systems engineers, to devise their own "intelligent" tools tailored to their specific requirements.

In addition to the ART family of networks, the Probabilistic Neural Network (PNN) [16] is another model suitable, in principle, for on-line, causal learning. The PNN algorithm is easy to use, simple to implement, and its training time is orders of magnitude faster than the back-propagation method [16]. Learning in the PNN is instantaneous in one-pass through the training samples. If sufficient data samples are available, the PNN is able to form complex non-linear decision surfaces that converge to the decision boundaries formulated by a Bayes classifier. More importantly, the decision boundaries can be modified incrementally when new samples are available, thus making the PNN suitable for on-line pattern classification problems. Evidently this leads to an increasing computational overhead as training progresses.

In the following sections, a hybrid neural network architecture that incorporates the PNN and a supervised ART model called Fuzzy ARTMAP (FAM) [14] is described. This network combines the incremental learning procedure of FAM and the probability estimation procedure of the PNN into a unified platform for on-line learning and prediction tasks. A number of modifications are introduced to enhance performance of the hybrid network. In addition, a multiple neural network architecture is introduced to aggregate outputs from several individual networks in order to increase the generalization and adaptability of the resulting system.

2 Fuzzy ARTMAP

FAM uses a mapping field to associate two unsupervised Fuzzy ART [12] networks together to form a supervised architecture. As depicted in Figure 1, FAM comprises two Fuzzy ART modules, ART_a and ART_b, coupled via a map field, F^{ab}. Following Carpenter, the variables in ART_a and ART_b are designated by superscripts a and b, respectively, and those in the map field are designated by a superscript ab. There are three layers of nodes in ART_a/ART_b: (i) F_0^a /F_0^b is a preprocessing layer that normalizes input patterns to be of equal norm; (ii) F_1^a /F_1^b is a comparison layer that receives both bottom-up and top-down patterns from F_0^a /F_0^b and F_2^a /F_2^b, respectively; (iii) F_2^a /F_2^b is a recognition layer that represents the category prototypes of input patterns. The map field, F^{ab}, is a single layer of nodes linking F_2^a and F_2^b.

During operation, ART_a receives an input pattern vector, $\boldsymbol{a}$, and ART_b receives the corresponding target vector, $\boldsymbol{b}$. These vectors are clustered into different categories in ART_a and ART_b, respectively. A mapping is then formed from ART_a to ART_b via the map field. The map field, however, does not directly associate $\boldsymbol{a}$ with $\boldsymbol{b}$. Instead, it associates the compressed category prototypes in F_2^a and F_2^b [13], [14]. A control mechanism is also included in the map field to detect predictive errors and to modulate supervised learning.

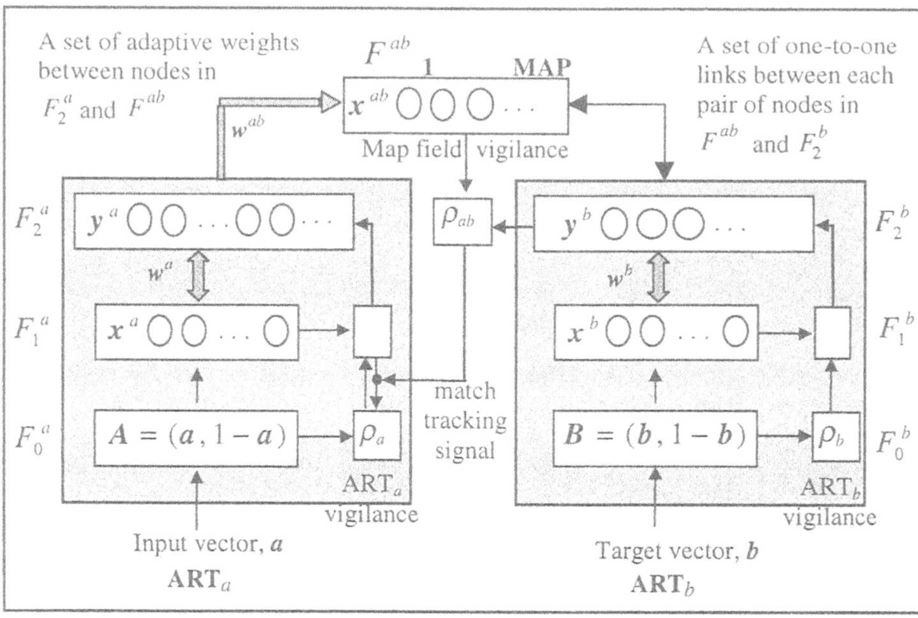

Figure 1. A schematic diagram of the Fuzzy ARTMAP architecture.

The learning algorithm of FAM can be divided into a few phases, as described below. For clarity, let $2M_a$ be the number of nodes in F_1^a, and N_a be the number of nodes in F_2^a. The output vectors, or the Short Term Memory (STM) activity, of F_1^a and F_2^a are denoted by $\boldsymbol{x}^a \equiv (x_1^a, \cdots, x_{2M_a}^a)$ and $\boldsymbol{y}^a \equiv (y_1^a, \cdots, y_{N_a}^a)$. The j-th ART_a weight vector, or the Long Term Memory (LTM) activity, is denoted by $\boldsymbol{w}_j^a \equiv (w_{j1}^a, \cdots, w_{j,2M_a}^a)$, $j = 1, \cdots, N_a$. All notation applies to ART_b when the superscripts or subscripts a and b are interchanged. In F^{ab}, the output vector is denoted by $\boldsymbol{x}^{ab} \equiv (x_1^{ab}, \cdots, x_{N_b}^{ab})$, and the weight vector from the j-th F_2^a node to F^{ab} is denoted by $\boldsymbol{w}_j^{ab} \equiv (w_{j1}^{ab}, \cdots, w_{j,N_b}^{ab})$, $j = 1, \cdots, N_a$.

2.1 Preprocessing and Initialization

All input vectors are preprocessed using a normalization technique called *complement-coding* [13], [14] that doubles the input dimension but maintains an equal norm for each vector. By using complement-coding, an M-dimensional input vector, $\boldsymbol{a}$, is normalized to a $2M$-dimensional vector, $\boldsymbol{A}$,

$$\boldsymbol{A} = (\boldsymbol{a}, \boldsymbol{a}^c) = (\boldsymbol{a}, 1-\boldsymbol{a}) \equiv (a_1, \cdots, a_M, 1-a_1, \cdots, 1-a_M) \quad (1)$$

The size of $\boldsymbol{A}$ is always made equal to M, *i.e.*,

$$|\boldsymbol{A}| = |(\boldsymbol{a}, \boldsymbol{a}^c)| = \sum_{i=1}^{M} a_i + (M - \sum_{i=1}^{M} a_i) = M \quad (2)$$

Thus, when complement-coding is used, the number of nodes in F_1^a is doubled.

At time $t = 0$, the weight vectors from each F_2^a node to the F_1^a layer are initialized to unity, *i.e.*,

$$w_{j1}^a(0) = \cdots = w_{j,2M_a}^a(0) = 1 \qquad j = 1, \cdots, N_a \quad (3)$$

This initialization indicates that each F_2^a node is *uncommitted* to any category prototype yet. When learning takes place, an F_2^a node becomes *committed* by modifying its weight vector to encode the input pattern. The resulting weight vector can be viewed as a *category prototype* of a cluster of input patterns.

From Figure 1, note that the number of nodes in F^{ab} is the same as the number of nodes in F_2^b, and there is a one-to-one permanent link between each corresponding pair of nodes. However, F_2^a is linked to F^{ab} through a set of adaptive weight vectors, $\boldsymbol{w}^{ab}$ Each component of $\boldsymbol{w}^{ab}$ is initialized to 1, *i.e.*, at time $t = 0$,

$$w_{j1}^{ab}(0) = \cdots = w_{j,N_b}^{ab}(0) = 1, \qquad j = 1, \cdots, N_a \quad (4)$$

This implies that before learning takes place, the j-th F_2^a prototype node is linked to every possible target output in F_2^b.

There are three important parameters in ART_a that govern the network dynamics [12], [14]: a choice parameter, $\alpha_a > 0$; a learning parameter, $\beta_a \in [0,1]$; and a vigilance parameter, $\rho_a \in [0,1]$. Setting the choice parameter to a small value slightly greater than zero (known as the conservative limit) ensures that the network conducts an efficient search within the existing category prototypes and minimizes recoding of similar prototypes [8]. The learning parameter decides the rate at which recoding should occur, whereas the vigilance parameter determines the degree of granularity of the learned categories. A low vigilance value leads to coarse category prototypes with broad generalization whereas a high vigilance value leads to fine category prototypes with narrow generalization.

2.2 Category Test and Search in Fuzzy ART

The input vector, $\boldsymbol{A}$, is first transmitted from F_1^a to F_2^a. The response of each F_2^a node is measured using a choice function

$$T_j(\boldsymbol{A}) = \frac{\left|\boldsymbol{A} \wedge \boldsymbol{w}_j^a\right|}{\alpha_a + \left|\boldsymbol{w}_j^a\right|} \tag{5}$$

The fuzzy AND/MIN operator ($\wedge$) is defined by [17]

$$(\boldsymbol{u} \wedge \boldsymbol{v})_i \equiv \min(u_i, v_i) \tag{6}$$

and the size, $|\cdot|$, is defined by

$$|\boldsymbol{u}| \equiv \sum_i |u_i| \tag{7}$$

i.e., the L_1 norm. The node that has the highest response, denoted as node J, is selected as the winning node,

$$T_J(\boldsymbol{A}) = \max\left\{T_j(\boldsymbol{A}) : j = 1, \cdots, N\right\} \tag{8}$$

If there is a tie in equation (8), then the category with the smallest index is chosen arbitrarily [14]. All other nodes $j \neq J$ are deactivated in accordance with the winner-take-all rule.

The winning node J propagates its weight vector back to F_1^a. A *vigilance test* is performed to measure the similarity between the transformed category prototype and the input vector against the vigilance parameter:

$$\frac{|\boldsymbol{x}^a|}{|\boldsymbol{A}|} = \frac{|\boldsymbol{A} \wedge \boldsymbol{w}_J^a|}{|\boldsymbol{A}|} \geq \rho_a \tag{9}$$

If the vigilance test is satisfied, resonance is said to occur and learning takes place, as defined below. However, if the test fails, node J is inhibited, *i.e.*, it is prohibited from participating in subsequent competitions. Input vector $\boldsymbol{A}$ is re-transmitted to F_1^a to search for a new winner. This process is repeated, consecutively disabling nodes in F_1^a, until either a category prototype is able to pass the vigilance test, or, if no such node exists, a new node is created to code the input vector.

2.3 Map-Field Prediction and Match-Tracking

The above test and search algorithm is conducted simultaneously at ART$_b$ until a winning node in F_2^b is found. Let node J be the winner in F_2^a and K be the winner in F_2^b. In accordance with the winner-take-all rule, the F_2^a and F_2^b output vectors, respectively, are set to

$$y_j^a = \begin{cases} 1 & \text{if } j = J \\ 0 & \text{otherwise} \end{cases} \quad j = 1, \cdots, N_a \tag{10}$$

$$y_k^b = \begin{cases} 1 & \text{if } k = K \\ 0 & \text{otherwise} \end{cases} \quad k = 1, \cdots, N_b \tag{11}$$

Node J sends a prediction signal to F^{ab} via $\boldsymbol{w}_J^{ab}$ whilst node K sends a confirmation signal to F^{ab} via the one-to-one link. An output vector, $\boldsymbol{x}^{ab}$, that obeys the following equation, is formed in F^{ab}:

$$\boldsymbol{x}^{ab} = \begin{cases} \boldsymbol{y}^b \wedge \boldsymbol{w}_J^{ab} & \text{if the } J\text{th } F_2^a \text{ node is active and } F_2^b \text{ is active} \\ \boldsymbol{w}_J^{ab} & \text{if the } J\text{th } F_2^a \text{ node is active and } F_2^b \text{ is inactive} \\ \boldsymbol{y}^b & \text{if } F_2^a \text{ is inactive and } F_2^b \text{ is active} \\ \boldsymbol{0} & \text{if } F_2^a \text{ is inactive and } F_2^b \text{ is inactive} \end{cases} \tag{12}$$

If it is assumed that F_2^a and F_2^b are always active, *i.e.*, every input pattern is associated with a target output in order for supervised learning to occur, then equation (12) reduces to

$$\boldsymbol{x}^{ab} = \boldsymbol{y}^b \wedge \boldsymbol{w}_J^{ab} \tag{13}$$

A map field vigilance test is then performed to confirm the predicted vector $\boldsymbol{x}^{ab}$ and the target vector $\boldsymbol{y}^b$, *i.e.*,

$$\frac{|\boldsymbol{x}^{ab}|}{|\boldsymbol{y}^b|} = \frac{|\boldsymbol{y}^b \wedge \boldsymbol{w}_J^{ab}|}{|\boldsymbol{y}^b|} \geq \rho_{ab} \tag{14}$$

where $\rho_{ab} \in [0,1]$ is a user-defined map field vigilance parameter. If the map field vigilance test is satisfied, learning ensues. However, failure of the vigilance test will trigger a series of corrective actions known as *match tracking* [13], [14].

At the beginning of each input vector presentation, the ART$_a$ vigilance parameter, ρ_a, equals a user-defined baseline vigilance, $\overline{\rho_a}$. In response to a mismatch in the map field vigilance test, ρ_a is raised to

$$\rho_a = \frac{|\boldsymbol{A} \wedge \boldsymbol{w}_J^a|}{|\boldsymbol{A}|} + \delta \tag{15}$$

where $0 < \delta << 1$. Therefore, the ART$_a$ vigilance test fails, *i.e.*, $|\boldsymbol{A} \wedge \boldsymbol{w}_J^a| < \rho_a\, |\boldsymbol{A}|$, and a new winner in F_2^a has to be found. In other words, match-tracking provides a means of selecting a prototype node (possibly a new, uncommitted node) which must satisfy both the ART$_a$ and map field vigilance tests (equations 9 and 14). If no such node exists, the input vector will be ignored [14].

2.4 Learning

Once search ends, learning takes place in each Fuzzy ART network as well as the map field. In ART$_a$, the winning weight vector, $\boldsymbol{w}_J^a$, is updated according to:

$$w_J^{a(new)} = \beta_a(A \wedge w_J^{a(old)}) + (1-\beta_a)w_J^{a(old)} \tag{16}$$

There are two learning modes: (i) *fast learning* corresponds to setting $\beta_a = 1$ for all time; (2) *fast-commit and slow-recode learning* corresponds to setting $\beta_a = 1$ for an uncommitted node and $\beta_a < 1$ for a committed node. The former rule allows the weight vector to converge to the asymptotic category boundary in one attempt, whereas the latter rule slowly varies the weight vector to make the system more resistant to noise [14]. The same learning rule is applicable in ART_b.

In map-field learning, an association is made between the active J-th node of F_2^a and the active K-th node of F_2^b. Learning is accomplished by updating w_J^{ab} to

$$w_{Jk}^{ab} = \begin{cases} 1 & \text{if } k = K \\ 0 & \text{if } k \neq K \end{cases} \tag{17}$$

In essence, equation (17) indicates that node J in a category prototype of target output K, and that the association is *permanent*. Therefore, in the absence of a target vector (for example, in the test phase), tracing the connection in w_J^{ab} can produce a prediction of the output.

2.5 Modifications to Fuzzy ARTMAP

In general, an independent Fuzzy ART network is employed as ART_b to self-organize the target vectors. However, in pattern classification problems where an input pattern belongs exclusively to one of the N possible target classes (one-from-N classification problems), ART_b can be replaced by a single layer containing N nodes. Then, the N-bit target pattern can be coded as a vector that has unit value corresponding to the actual output class and zero for all others.

From the viewpoint of statistical pattern classification, estimates of the probabilities that an input pattern belongs to more than one target class with different degrees of class membership are often required. While the probability of having two or more identical, real-valued input vectors in a data set is vanishingly small, in cases of discrete-valued patterns, especially binary-valued patterns of low dimension, the same input vector may be associated with distinct outputs with different probabilities. Therefore, it is advantageous if the category prototypes in F_2^a can be linked to more than one target class in F_2^b, *i.e.*, a *one-to-many mapping* between ART$_a$ and ART$_b$.

Since the original FAM algorithm is unable to establish such a mapping, a modification has been proposed to implement one-to-many mapping [18]. Thus, we proposed to impose a constraint on ρ_a during match-tracking, as follows

$$0 \le \rho_a \le \min\left(1, \frac{|A \wedge w_J^a|}{|A|} + \delta\right) \tag{18}$$

If there is a perfect match in the vigilance test (equation 9) between the winning F_2^a prototype and the input vector but subsequently a failure in the map field vigilance test (a predictive error as in equation 14), the above constraint will lead to the selection of a new node in F_2^a, or another perfectly matched prototype to encode the input vector, instead of ignoring it as would have occurred according to the original FAM algorithm. Hence, a one-to-many mapping is established in the map field to associate identical prototypes with more than one target class.

3 The Probabilistic Neural Network

The PNN realizes the Parzen-window probability estimation procedure [19] in a feedforward neural network architecture. Parzen devised a non-parametric approach to estimate the probability density function (pdf) from a set of data samples. Let $x_1, \cdots, x_n$ denote a set of one-dimensional, statistically independent random variables with an unknown pdf, $f(x)$. Using the Parzen-window technique, $f(x)$ can be approximated by $\hat{f}(x)$ as:

$$\hat{f}(x) = \frac{1}{n}\sum_{i=1}^{n}\frac{1}{\lambda_n}K\left(\frac{x - x_i}{\lambda_n}\right) \tag{19}$$

where $K(\cdot)$ is a kernel function (sometimes known as a potential or weighting function) centered on each data point, and λ_n is a smoothing parameter dependent on the sample size, n. In [20], Cacoullos extended Parzen's results to cover multi-dimensional cases. Assuming that a Gaussian kernel function is used, the multivariate pdf estimate for a set of M-dimensional vectors of random variables, $\boldsymbol{x}_1, \cdots, \boldsymbol{x}_n$, can be expressed as

$$\hat{f}(\boldsymbol{x}) = \frac{1}{N(2\pi)^{M/2}\sigma^M} \sum_{i=1}^{N} \exp\left(-\frac{(\boldsymbol{x}-\boldsymbol{x}_i)^t(\boldsymbol{x}-\boldsymbol{x}_i)}{2\sigma^2} \right) \tag{20}$$

where σ is the smoothing parameter of the kernel function. The estimated pdf can be viewed as a weighted average of a set of multivariate Gaussian functions, one for each data sample. The choice of σ has a profound effect on the estimated function. If σ is too small, each data point will exert a distinct mode corresponding to its location in the input space on the resultant pdf. If σ is too large, the estimated function could achieve a greater degree of interpolation at the expense of losing information on local properties. Figure 2 depicts an example illustrating the effect of different values of σ on the estimated pdf.

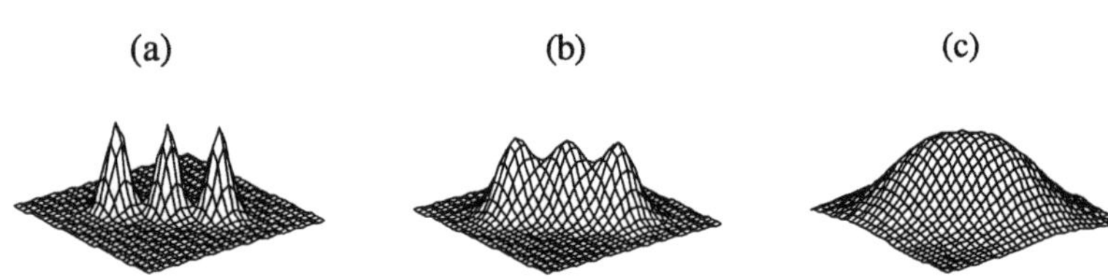

Figure 2. Effects of the smoothing parameter, σ, on the Gaussian kernel function: (a) a small value of σ; (b) a medium value of σ; (c) a large value of σ.

The Parzen probability estimation procedure mentioned above can be implemented using the PNN. Figure 3 shows that the PNN has four layers of nodes, namely the input layer, pattern layer, summation layer, and output layer. The input layer receives an M-dimensional vector, $\boldsymbol{x}$, and propagates to all the nodes in the pattern layer. The nodes in the pattern layer are grouped into different categories according to their target classes. Assuming that all vectors are normalized to unit length, the exponential part in equation (20) can be computed as follows:

$$z_i = \boldsymbol{x} \cdot \boldsymbol{w}_i \tag{21}$$

$$g(z_i) = \exp(\frac{z_i - 1}{\sigma^2}) \tag{22}$$

where $\boldsymbol{w}_i$ is the weight vector of the i-th pattern node, and $g(z_i)$ represents the Gaussian kernel.

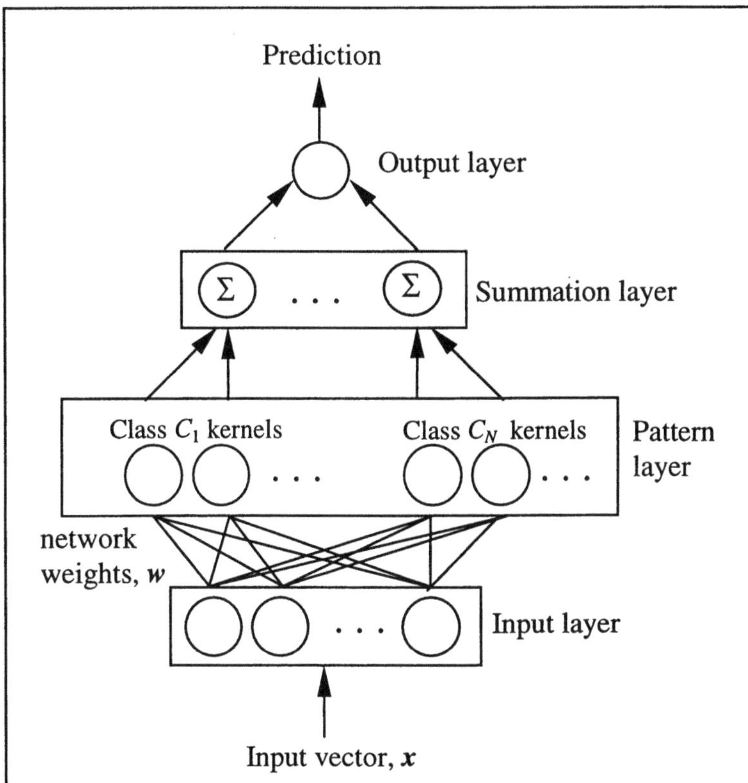

Figure 3. A schematic diagram of the Probabilistic Neural Network.

The summation nodes are used to sum outputs from the pattern nodes to yield the estimated conditional pdf of each of the k classes, $p(\boldsymbol{x}|C_k)$. Different decision criteria can then be applied at the output layer for producing a final prediction. For example, the risk-weighted Bayesian posterior probabilities can be implemented as follows

$$p(C_k \mid \boldsymbol{x}) = p(\boldsymbol{x} \mid C_k)p(C_k)l(C_{jk}) \tag{23}$$

where the conditional pdf, $p(\mathbf{x}|C_k)$, is weighted by the prior probability, $p(C_k)$, and the risk or loss factor, $l(C_{jk})$ – the risk of choosing C_k when C_j is the actual class, $j \neq k$. While the prior probabilities can be estimated from the relative frequency of samples in a data set, the loss factors have to be determined subjectively to reflect the significance of incorrect decisions. If there is no particular reason to penalize misclassifications differently by using different loss factors, then the class that has the highest posterior probability should be selected as the predicted output.

The learning process in the PNN is non-iterative and is orders of magnitude faster than the MLP trained with backpropagation [16]. It is accomplished by generating a pattern node, connecting it to the summation node of the target class, and assigning the input vector as the weight vector. Thus, N summation nodes will be created, one for each target class, if there are N different classes. Although the training procedure is very fast, one of its drawbacks is that the number of pattern nodes required in the PNN is the same as the number of training samples. While this is not an issue for small training sets, it will result in an explosive number of pattern nodes if large or unbounded data sets are employed. Besides, the storage requirement and computational burden increase correspondingly with the size of the data set.

4 Probabilistic Fuzzy ARTMAP

Pattern classification problems have been widely studied using statistical approaches such as probability density estimation, discriminant analysis, and Bayes decision theory [21], [22]. The Bayes decision theory has been generally accepted to yield minimum misclassification rates in a statistical sense. FAM, however, is a deterministic system, *i.e.*, it classifies an input pattern to only one target class without providing any probability estimation of the output. Therefore, it is useful to investigate how to incorporate statistical principles into the FAM algorithm. On the other hand, we know that the drawback of the PNN is the need to recruit a new pattern node for each input vector, resulting in an unwieldy network structure for large data sets, *i.e.*, there is no code compression. This problem can be alleviated by using a *clustering technique* to find a set of reference vectors, or prototypes, to represent data samples that are closely clustered in the input space. Instead of using all input vectors, these prototypes are employed to estimate the pdfs. Consequently, the computational and storage requirements are reduced, while the capability of estimating pdf is maintained.

It thus appears that the learning methodologies of FAM and the PNN complement one another, *i.e.*, FAM provides the clustering ability while the PNN provides the probability estimates. Indeed, there is a close similarity between the network structures of FAM and the PNN, as depicted in Figure 4. The F_1^a and F_2^a layers of FAM correspond to the input and pattern layers of the PNN, respectively, whereas the map field layer, F^{ab}, corresponds to the summation layer. In essence, in one-from-N classification, each node in F_2^a is permanently associated with only one node in F^{ab} which is then linked to the target output in F_2^b. Thus, nodes in F^{ab} can be used to sum outputs from all prototype F_2^a nodes that are connected to their respective target classes, taking the role of the summation nodes in the PNN. Subsequently, F_2^b can function as the output layer of the PNN in which the most active node is selected to produce a prediction.

In view of the complementary characteristics between FAM and the PNN, a *Probabilistic Fuzzy ARTMAP (PFAM)* network [23] has been proposed, in which both the networks are integrated into a unified platform for on-line classification and prediction tasks. The advantage of the synergism is two-fold: (i) FAM is used as the underlying clustering algorithm to reduce the number of pattern nodes as required in the PNN, (ii) the PNN is used as the probability estimation algorithm to provide probabilistic predictions for FAM. The resulting hybrid network is able to classify an arbitrary sequence of input patterns into different categories, and to give a probabilistic estimate of the predicted classes for facilitating the application of the Bayes decision rule. In summary, the PFAM algorithm is divided into two phases:

(i) Learning Phase

FAM is utilized as the underlying, supervised, clustering algorithm to categorize the input patterns into a set of prototypes. The weight vectors of each F_2^a node encode these cluster prototypes.

(ii) Prediction Phase

The PNN is utilized to estimate the probability of the input vector belonging to each output class. The Bayes decision rule can then be applied to select the maximum *a posteriori* output, or to implement other risk-weighted rules.

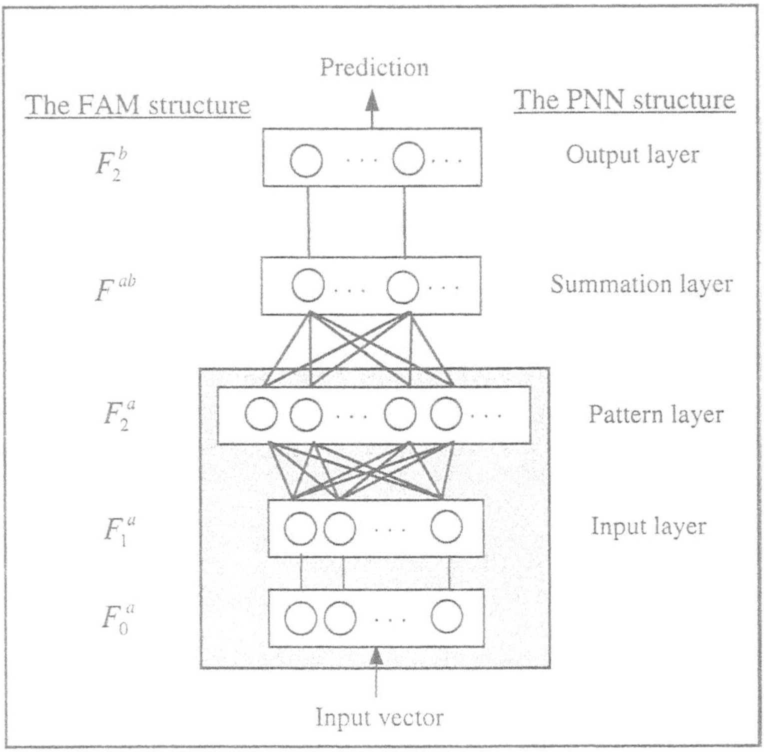

Figure 4. A schematic diagram of the similarity between the network structures of FAM and the PNN.

A number of modifications to the network dynamics are needed for effective combination of FAM and the PNN and to improve the generalization of the resulting system, as described in the following sections.

4.1 Map-Field Dynamics

In FAM, only one of the components in the map field weight vectors, w_j^{ab}, $j = 1,\cdots, N_a$, has unit value that indicates the association between category prototypes in F_2^a with their target outputs in F_2^b. However, to estimate the prior class probabilities, the connection between the F_2^a and F^{ab} nodes has to be increased correspondingly, *i.e.*, if the j-th F_2^a node successfully predicts a target output, the link in $\boldsymbol{w}_j^{ab}$ should be incremented by one to keep track of the number of samples represented by each prototype node. In addition, estimates from the kernel functions can be weighted by the strength of $\boldsymbol{w}_j^{ab}$.

In PFAM, during the map field prediction cycle, $\boldsymbol{x}^{ab}$ in equation (13) is replaced by

$$\boldsymbol{x}^{ab} = \boldsymbol{y}^b \wedge \frac{\boldsymbol{w}_J^{ab}}{\left|\boldsymbol{w}_J^{ab}\right|} \tag{24}$$

When learning takes place, the map field weight vector is updated according to

$$\boldsymbol{w}_J^{ab} = \boldsymbol{w}_J^{ab} + \boldsymbol{x}^{ab} \tag{25}$$

Subsequently, the number of samples coded by the prototype nodes in F_2^a that belongs to the k–th target class is

$$S_k = \sum_{j=1}^{N_a} w_{jk}^{ab} \tag{26}$$

and the total number of samples coded by the network is

$$S_{Total} = \sum_{k=1}^{N_b} S_k \tag{27}$$

From equations (26) and (27), the prior probability of class k, $p(C_k)$, can be calculated using

$$p(C_k) = \frac{S_k}{S_{Total}} \tag{28}$$

4.2 Center Estimation

One of the difficulties of the Parzen-window density estimation is the selection of a suitable kernel function and its associated parameters. Here, the Gaussian kernel function has been chosen because it exhibits properties that are mathematically and computationally tractable [24], [25]. There are two important parameters, *i.e.*, the mean and covariance corresponding to the center and width of the Gaussian function, that need to be estimated.

With complement coding and the fast learning rule in FAM, the prototype nodes are represented by the vertices of hyper-rectangles that define boundary of input clusters [14]. In other words, they do not reflect the centroids of input clusters. Figure 5 depicts two possible category structures of different classes in a two-dimensional example.

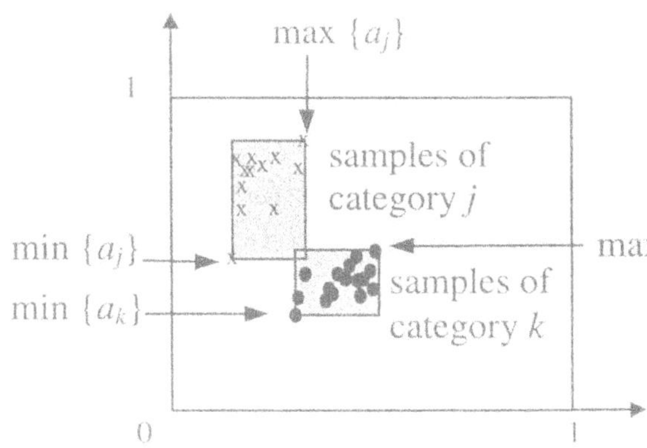

Figure 5. In a two-dimensional example, the category structures are coded by vertices of rectangles, and thus do not reflect the centroids of pattern clusters represented by the category prototypes.

To overcome this problem, a new set of weight vectors is introduced in PFAM and is placed in the ART$_a$ module. They are called the *center weight vectors*, *i.e.*,

$$\boldsymbol{w}_j^{a-c} \equiv (w_{j1}^{a-c},\cdots,w_{j,M_a}^{a-c}) \quad j=1,\cdots,N_a \tag{29}$$

Notice that $\boldsymbol{w}_j^{a-c}$ only covers the original dimension of the input vector, *i.e.*, M_a. Thus, in PFAM the original weight vectors, $\boldsymbol{w}_j^a$, (hereafter called *category weight vectors*) are used for the categorization of input patterns into different classes in the learning phase. In the prediction phase, only the first half of the number of F_1^a nodes is activated in which the original input vector, $\boldsymbol{a}$, and $\boldsymbol{w}_j^{a-c}$ are propagated to F_2^a for the pdf estimation.

Initially, all center weight vectors are set to

$$\boldsymbol{w}_j^{a-c} \equiv (w_{j1}^{a-c},\cdots,w_{j,M_a}^{a-c}) = (0.0,\cdots,0.0) \quad j=1,\cdots,N_a \tag{30}$$

When learning takes place, the center weight vector of the J-th F_2^a winning node is updated according to:

$$\left(\boldsymbol{w}_J^{a-c}\right)^{\text{new}} = \left(\boldsymbol{w}_J^{a-c}\right)^{\text{old}} + \frac{1}{\left|\boldsymbol{w}_J^{ab}\right|}\left(\boldsymbol{a} - \left(\boldsymbol{w}_J^{a-c}\right)^{\text{old}}\right) \tag{31}$$

Equation (31) is equivalent to a recursive mean computation for estimating the centroid of the J-th prototype node, bounded by its category weight vector, as shown in Figure 6. The size $|\boldsymbol{w}_j^{ab}|$ indicates the number of input patterns that has been coded by node J.

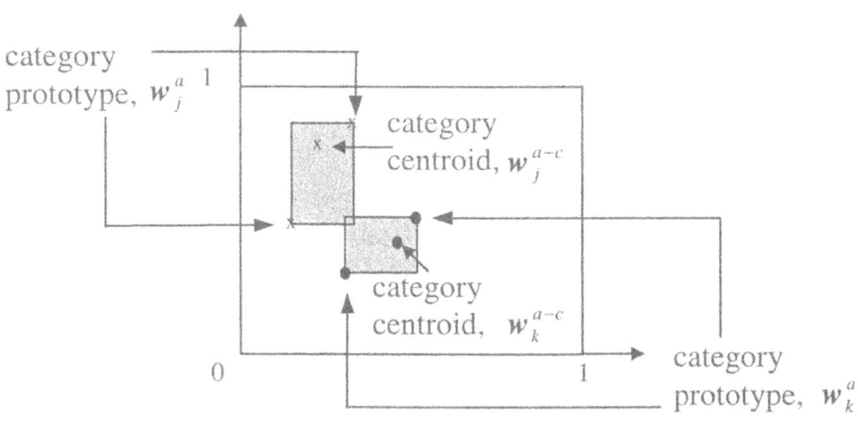

Figure 6. In addition to the min-max category weight vectors, a recursive mean algorithm is used to estimate the center weight vector of each category.

4.3 Width Estimation

In PFAM, if a general Gaussian kernel function is to be used, *i.e.*,

$$g\left(\left\|a - w_j^{a-c}\right\|\right) = \frac{1}{(2\pi)^{M/2}\left|\Sigma_j\right|^{1/2}} \exp\left(-\frac{1}{2}(a - w_j^{a-c})^t \Sigma_i^{-1} (a - w_j^{a-c})\right) \quad (32)$$

then it is necessary to estimate Σ_j (the covariance matrix of the j-th F_2^a prototype node). In the PNN, a global smoothing parameter, σ, is used for all the kernels. This means that all the Gaussian functions have the same covariance matrix; the covariance matrix is diagonal; and the eigenvalues are equal, such that $\Sigma_j = \sigma^2 \boldsymbol{I}$. As suggested in [26], it is important for the kernel functions to maintain local properties in certain classification tasks. While there is no problem if more than one pdf belonging to the same class overlap each other, pdfs of different classes have to be separated so that they can exhibit their individual decision boundaries. The use of a global smoothing parameter may not, therefore, yield such behavior.

Instead of using a global smoothing parameter, each kernel in PFAM has its own smoothing parameter. In line with the rationale to limit the kernel width to the location of the nearest neighbor of a distinct class, the width of the j-th prototype node can be estimated using

$$\sigma_j = \frac{1}{r} \min \left\| w_j^{a-c} - w_k^{a-c} \right\| \quad \text{class}\, j \neq k \quad (33)$$

where r is an "overlapping parameter". Equation (33) yields an enclosing hypersphere around each prototype node. The role of r is to obtain a smooth pdf interpolated from distant kernels. Since the covariance matrix of each kernel is now simplified to $\Sigma_j = \sigma_j^2 \boldsymbol{I}$, calculation of the determinant and inversion of the covariance matrix as in equation (32) is avoided. As a result, equation (32) is reduced to

$$g\left(\left\|a - w_j^{a-c}\right\|\right) = \frac{1}{(2\pi)^{M/2}\sigma_j^M} \exp\left(-\frac{(a - w_j^{a-c})^t (a - w_j^{a-c})}{2\sigma_j^2}\right) \quad (34)$$

To ensure that the pdfs constructed are based on the most up-to-date information stored in PFAM, parameters of the Gaussian kernels have to be re-estimated whenever a new sample is learned by the network. The above width estimation procedure, therefore, provides an alternative to preserve local neighborhood of input patterns and, at the same time, to reduce the computational and storage requirement of the network.

The use of the nearest neighbor (NN) heuristic in equation (33) sometimes shows signs of "conservativeness" by forming kernels with small width in order to prevent a large amount of overlapping among different class prototypes [26]. The resulting kernel functions would have a relatively peaked maximum, and they fail to interpolate smoothly by giving undue emphasis to individual prototypes. One way to mitigate this problem is to choose a suitable value of r in equation (33). Another way is to use a K nearest neighbor (K-NN) heuristic to estimate σ_j. Instead of restricting σ_j to the nearest distinct neighbor, the width of the j-th F_2^a prototype node can be estimated by averaging distances from K distinct neighbors, *i.e.*,

$$\sigma_j = \frac{1}{K}\sum_{k=1}^{K}\left\|w_j^{a-c} - w_k^{a-c}\right\| \quad \text{class}\, j \neq k,\ 1 \leq K \leq N_a - 1 \tag{35}$$

When $K = 1$, equation (35) reduces to equation (33) with $r = 1$.

4.4 Summary of the PFAM Algorithm

The algorithm of PFAM for on-line learning and probability estimation is summarized as follows.

(a) Learning Phase:

(i) Complement code an M-dimensional input vector, $\boldsymbol{a} \in [0,1]^M$, in F_0^a to a $2M$-dimensional vector $\boldsymbol{A} = (\boldsymbol{a}, 1-\boldsymbol{a})$ in F_1^a.

(ii) Feed forward $\boldsymbol{A}$ from F_1^a to F_2^a via $\boldsymbol{w}_j^a$. Compute the match function as in equation (5) and select the winning node (denoted as node J).

(iii) Feed back $\boldsymbol{w}_J^a$ from F_2^a to F_1^a and perform the vigilance test as in equation (9).

(iv) If the vigilance test fails, then inhibit the winning node and go to step (ii).

(v) Send a prediction from F_2^a to F^{ab} using equation (24) and perform the map field vigilance test as in equation (14).

(vi) If the map field vigilance test fails, then inhibit the winning node and trigger match-tracking with implementation of equation (18), and go to step (ii).

(vii) Learning: update the weight vectors $\boldsymbol{w}_J^a$, $\boldsymbol{w}_J^{ab}$ and $\boldsymbol{w}_J^{a-c}$ according to equations (16), (25), and (31) respectively.

(b) Prediction Phase:

(i) Feed forward the original input vector, $\boldsymbol{a}$, from F_0^a to F_1^a, and then to F_2^a together with $\boldsymbol{w}_j^{a-c}$.

(ii) For each j-th F_2^a node, compute the width estimate using equation (33) or (35), and compute the kernel activation function as in equation (34).

(iii) Sum the kernel estimates from F_2^a in the form of equation (20) to the corresponding target classes in the map field to obtain the pdfs, $p(\boldsymbol{x}|C_k)$.

(iv) Compute the prior probabilities, $p(C_k)$, using equations (26) to (28), and compute the weighted posterior probabilities, $p(C_k|\boldsymbol{x})$, using equation (23).

(v) Select the highest posterior estimate or the minimum-risk estimate according to the Bayes rule and output the predicted target class at ART_b.

5 Multiple Classifier Systems

The use of a single system for pattern classification hinges on the assumption that the system is able to capture and process all the input features regardless of what the features might be. In cases where the above assumption fails to hold true, *e.g.*, the input features might consist of a mixture of syntactic primitives, linguistic variables, continuous, discrete, or nominal attributes, presenting all these features to one classifier for it to make a decision is often difficult and can result in poor performance. Furthermore, concatenating all the features into a high-dimensional input vector will unduly induce the problem known as the "curse-of-dimensionality" [22]. Hence, many researchers have proposed the application of a multiple classifier system (MCS) to process the same data set, and the combination of results using some algorithm to reach a combined decision [27]-[29].

From the perspective of FAM and PFAM, there is another reason for the use of multiple classifiers, *i.e.*, to minimize the effect of data ordering. As pointed out in [14], the formation of ART_a cluster prototypes is affected by the sequence of input sample presentations. This would subsequently lead to different predictions of target classes, and thus different accuracy scores for each realization of FAM. Although PFAM adopts a distributed processing approach where all prototypes are used to reach a prediction (instead of the "winner-take-all" rule as in FAM), it still does not escape from this problem – good long-run performance of the network depends on a good initial formation of cluster prototypes. The data ordering effect is further exacerbated if the prototypes are to be established autonomously, on-line, because in this case the input samples are presented only once, and in a fixed order. One way to mitigate this problem is to train a pool of networks off-line, each with a different ordering of input samples. During the prediction phase, the results from several networks can be combined to give an overall prediction. It is therefore worthwhile to investigate how to integrate decisions from multiple PFAM networks so that a robust and high performance classification system is realized in real applications.

Here, three decision combination algorithms are discussed to form a PFAM-based MCS, as shown in Figure 7. The first one is a simple majority-voting scheme where the target class that receives the highest number of votes is selected as the final prediction. The second method is based on the Bayesian theorem that is commonly employed in evidence gathering and uncertainty reasoning. The third one is the so-called behavior-knowledge space approach proposed in [29].

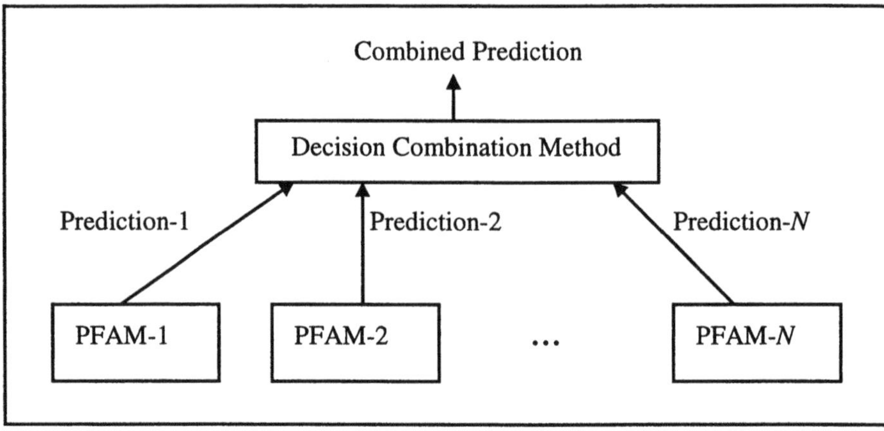

Figure 7. A schematic diagram of a multiple classifier system. There are N modules of independent PFAM classifier. Outputs from all classifiers are combined using some decision combination algorithm to give an overall prediction.

5.1 Voting

Suppose there are M target classes where each class is represented by C_i, $\forall i \in \Lambda = \{1, 2, \cdots, M\}$, the task of a classifier is to assign the input sample, $\boldsymbol{x}$, to one of the $M + 1$ classes, with the $(M+1)$-th class denoting that the classifier rejects $\boldsymbol{x}$. In a MCS, all classifiers are not compelled to predict the same class. A common and simple way for resolving conflicts in opinions in human social life is by voting. The same principle can be adopted to reach a final outcome for the predictions from multiple classifiers. The most conservative method is the unanimous voting strategy where $\boldsymbol{x}$ is classified to C_i, if and only if all the classifiers predict C_i, otherwise $\boldsymbol{x}$ is rejected. Another strategy is the voting-by-majority rule for determining the final result, as presented below.

Given an input sample, $\boldsymbol{x}$, if there are K classifiers denoted by $e_1, \cdots, e_K$, the problem is to produce a combined result, $E(\boldsymbol{x}) = j, j \in \Lambda \cup \{M + 1\}$, from all the K predicted classes, $e_k(\boldsymbol{x}) = j_k$, $k = 1, \cdots, K$. A binary function can be used to represent the number of votes, *i.e.*,

$$V_k(\boldsymbol{x} \in C_i) = \begin{cases} 1, & \text{if } e_k(\boldsymbol{x}) = i, i \in \Lambda \\ 0, & \text{otherwise} \end{cases} \tag{36}$$

Then, by summing the votes from all K classifiers for each C_i,

$$V_E(\boldsymbol{x} \in C_i) = \sum_{k=1}^{K} V_k(x \in C_i), \quad i = 1, \cdots, M \tag{37}$$

the combined result, $E(\boldsymbol{x})$, can be determined by

$$E(\boldsymbol{x}) = \begin{cases} j, & \text{if } V_E(\boldsymbol{x} \in C_j) = \max_{i \in \Lambda} V_E(\boldsymbol{x} \in C_i) \\ & \text{and } \dfrac{V_E(\boldsymbol{x} \in C_j)}{K} \geq \lambda \\ M+1, & \text{otherwise} \end{cases} \tag{38}$$

where $0 \leq \lambda \leq 1$ is a user-defined threshold that controls the confidence in the final decision.

5.2 The Bayesian Approach

The voting strategy is solely based on the predicted outcomes produced from all classifiers, where each classifier is treated equally without considering its errors. A more reasonable approach is to take into account the predictive accuracy of each classifier. The predictions from highly accurate classifiers should be given more weight than those from the less accurate ones. This is the rationale behind the use of the Bayesian approach to combine decisions from multiple classifiers. The Bayesian combination algorithm presented below is proposed in [27], but the basic idea has been used previously for evidence propagation and uncertainty reasoning in intelligent systems [30].

Given a data set containing N samples, the performance (correct and incorrect predictions) of the k-th classifier, e_k, is recorded in its confusion matrix constructed as follows:

$$CM^k = \begin{pmatrix} n_{11}^k & n_{12}^k & \cdots & n_{1(M+1)}^k \\ n_{21}^k & n_{22}^k & \cdots & n_{2(M+1)}^k \\ \vdots & \vdots & \ddots & \vdots \\ n_{M1}^k & n_{M2}^k & \cdots & n_{M(M+1)}^k \end{pmatrix} \tag{39}$$

where n_{ij}^k, $i = 1,\cdots, M$, $j = 1,\cdots, M + 1$ indicates the number of samples belonging to C_i, but assigned to class j by e_k. From the confusion matrix, the total number of samples encountered by e_k is

$$N = \sum_{i=1}^{M} \sum_{j=1}^{M+1} n_{ij}^k \tag{40}$$

and the number of samples belonging to C_i is

$$n_{i\bullet}^k = \sum_{j=1}^{M+1} n_{ij}^k \tag{41}$$

i.e., summation through row i. The number of samples that is assigned to class j by e_k is

$$n_{\bullet j}^k = \sum_{i=1}^{M} n_{ij}^k \tag{42}$$

i.e., summation through column j. This confusion matrix provides information regarding a classifier's ability to classify accurately samples from a particular target class. In the event $e_k(\boldsymbol{x}) = j$ (classifier e_k predicts that $\boldsymbol{x}$ belongs to C_j), the truth that $\boldsymbol{x}$ really comes from class C_j is associated with a factor of uncertainty. By utilizing information stored in the confusion matrix, the uncertainty of proposition $\boldsymbol{x} \in C_i$ given $e_k(\boldsymbol{x}) = j$ can be computed according to

$$P(\boldsymbol{x} \in C_i \mid e_k(\boldsymbol{x}) = j) = \frac{n_{ij}^k}{n_{\bullet j}^k} = \frac{n_{ij}^k}{\sum_{i=1}^{M} n_{ij}^k}, \quad i = 1,\cdots, M \tag{43}$$

From the viewpoint of uncertainty reasoning, the confusion matrix of a classifier can be regarded as a collection of evidence supporting different target classes, and equation (43) can be interpreted as a set of belief functions, $bel(\cdot)$, on M propositions that $\boldsymbol{x} \in C_i$ [30]. The higher the belief function of a proposition is, the more likely that it is true. With K classifiers, there will be K confusion matrices and K events, $e_1(\boldsymbol{x}) = j_1,\cdots, e_K(\boldsymbol{x}) = j_K$. Each classifier e_k expresses its belief functions as:

$$\begin{aligned} &bel(\boldsymbol{x} \in C_i \mid e_k(\boldsymbol{x}) = j_k, EN) \\ &= P(\boldsymbol{x} \in C_i \mid e_k(\boldsymbol{x}) = j_k) \qquad i = 1, \cdots, M \end{aligned} \tag{44}$$

where EN denotes the common classification environment that consists of all K events. The problem now is how to integrate the belief functions from K classifiers into a combined set of belief functions.

In [30], a Bayesian formalism is adopted to propagate and update the belief functions in the Bayesian network. The same idea has been used in [27] to combine K sets of belief function. Using Bayes' theorem, equation (44) is expanded to

$$\begin{aligned} bel(i) &= bel(\boldsymbol{x} \in C_i | e_1(\boldsymbol{x}) = j_1, \cdots, e_K(\boldsymbol{x}) = j_K, EN) \\ &= P(\boldsymbol{x} \in C_i | e_1(\boldsymbol{x}) = j_1, \cdots, e_K(\boldsymbol{x}) = j_K, EN) \\ &= \frac{P(e_1(\boldsymbol{x}) = j_1, \cdots, e_K(\boldsymbol{x}) = j_K | \boldsymbol{x} \in C_i, EN) P(\boldsymbol{x} \in C_i | EN)}{P(e_1(\boldsymbol{x}) = j_1, \cdots, e_K(\boldsymbol{x}) = j_K | EN)} \end{aligned} \tag{45}$$

In order to simplify the combination of the belief functions, it is assumed that the environment EN consists of K independent events with M mutually exclusive sets of target output. Thus, the joint probability is reduced to

$$\begin{aligned} &\frac{P(e_1(\boldsymbol{x}) = j_1, \cdots, e_K(\boldsymbol{x}) = j_K \mid \boldsymbol{x} \in C_i, EN)}{P(e_1(\boldsymbol{x}) = j_1, \cdots, e_K(\boldsymbol{x}) = j_K \mid EN)} \\ &= \frac{\prod_{k=1}^{K} P(e_k(\boldsymbol{x}) = j_k \mid \boldsymbol{x} \in C_i, EN)}{\prod_{k=1}^{K} P(e_k(\boldsymbol{x}) = j_k \mid EN)} \end{aligned} \tag{46}$$

Using the Bayes' rule

$$\frac{P(e_k(\boldsymbol{x}) = j_k | \boldsymbol{x} \in C_i, EN)}{P(e_k(\boldsymbol{x}) = j_k | EN)} = \frac{P(\boldsymbol{x} \in C_i | e_k(\boldsymbol{x}) = j_k, EN)}{P(\boldsymbol{x} \in C_i | EN)} \tag{47}$$

equation (46) becomes

$$\frac{\prod_{k=1}^{K} P(e_k(\boldsymbol{x}) = j_k | \boldsymbol{x} \in C_i, EN)}{\prod_{k=1}^{K} P(e_k(\boldsymbol{x}) = j_k | EN)} = \frac{\prod_{k=1}^{K} P(\boldsymbol{x} \in C_i | e_k(\boldsymbol{x}) = j_k, EN)}{\prod_{k=1}^{K} P(\boldsymbol{x} \in C_i | EN)} \tag{48}$$

Substituting (48) into (45) gives

$$bel(i) = \frac{\prod_{k=1}^{K} P(\boldsymbol{x} \in C_i | e_k(\boldsymbol{x}) = j_k, EN)}{\prod_{k=1}^{K} P(\boldsymbol{x} \in C_i | EN)} P(\boldsymbol{x} \in C_i | EN) \tag{49}$$

To further simplify calculation of the combined belief function, the following estimate can be used [27]

$$bel(i) = \frac{\prod_{k=1}^{K} P(\boldsymbol{x} \in C_i | e_k(\boldsymbol{x}) = j_k)}{\sum_{i=1}^{M} \prod_{k=1}^{K} P(\boldsymbol{x} \in C_i | e_k(\boldsymbol{x}) = j_k)} \tag{50}$$

where each $P(\boldsymbol{x} \in C_i \mid e_k(\boldsymbol{x}) = j_k)$ can be computed from the confusion matrix using equation (43) by replacing j with j_k. Based on the combined belief functions, the one with the highest estimate is selected as the final outcome, *i.e.*,

$$E(\boldsymbol{x}) = \begin{cases} j, & \text{if } bel(j) = \max_{i \in \Lambda} bel(i) \text{ and } bel(j) \geq \lambda \\ M+1, & \text{otherwise} \end{cases} \tag{51}$$

Again, $0 \leq \lambda \leq 1$ is a user-defined threshold to regulate confidence associated with the final decision.

5.3 The Behavior-Knowledge Space Approach

One of the criticisms of the Bayesian approach is the assumption that all classifiers must operate independently in order to tackle the computation of the joint probabilities. This assumption is unlikely to hold in many applications. To avoid using the assumption, a method called the Behavior-Knowledge Space (BKS) that concurrently records the decisions of all classifiers on each input sample is proposed in [29].

A BKS is a K-dimensional space in which each dimension corresponds to the decision of one classifier. The intersection of the decision from multiple classifiers occupies one unit of the BKS, *e.g.*, BKS($e_1(\boldsymbol{x}) = j_1, \cdots, e_K(\boldsymbol{x}) = j_K$) denotes a BKS unit where each e_k produces a prediction j_k, $k = 1,\cdots, K$. In each BKS unit, there are M partitions (cells) for accumulating the number of samples actually belonging to C_i. As an example, suppose two classifiers are used to categorize the input samples into M target classes. Then a two-dimensional BKS can be formed as shown in Figure 8.

Each BKS unit, U_{ij}, is further divided into M cells, $n_1^H,\cdots, n_M^H$, where H denotes the overall prediction $e_1(\boldsymbol{x}) = j_1, \cdots, e_K(\boldsymbol{x}) = j_K$. The frequency of the number of samples belonging to C_i is recorded in each n_i^H, $i = 1,\cdots, M$. When an input sample, $\boldsymbol{x}$, is presented, one of the BKS units will become active (known as the focal unit) after the decisions from all K classifiers have been received. In the above example, U_{34} will be selected as the focal unit if $e_1(\boldsymbol{x}) = 3$ and $e_2(\boldsymbol{x}) = 4$. Then, the total number of samples in the focal unit is computed:

$$T(H) = \sum_{i=1}^{M} n_i^H \tag{52}$$

and the best representative class (*i.e.*, the one that contains the highest number of samples) is identified,

$$R(H) = j \ \text{ where } n_j^H = \max_{i\in\Lambda}(n_i^H) \tag{53}$$

The decision rule for determining the final outcome is

$$E(\boldsymbol{x}) = \begin{cases} R(H), & \text{if } T(H) > 0 \text{ and } \dfrac{n_{R(H)}^H}{T(H)} \geq \lambda \\ M+1, & \text{otherwise} \end{cases} \tag{54}$$

where $0 \leq \lambda \leq 1$ is a user-defined confidence threshold.

e_2 \ e_1	1	2	$\cdots$	$M+1$
1	U_{11}	U_{12}	$\cdots$	$U_{1(M+1)}$
2	U_{21}	U_{22}	$\cdots$	$U_{2(M+1)}$
$\vdots$	$\vdots$	$\vdots$	$\ddots$	$\vdots$
$M+1$	$U_{(M+1)1}$	$U_{(M+1)2}$	$\cdots$	$U_{(M+1)(M+1)}$

Figure 8. A two-dimensional Behavior-Knowledge Space.

The BKS appears to be similar to the confusion matrix used in the Bayesian approach. However, the Bayesian method needs to compute the multiplication of evidence taken from the confusion matrices to approximate the joint probability of K events when evaluating the combined belief function. This step is avoided in the BKS approach, in which a final decision is made by assigning the input sample directly to the class that has accumulated the largest number of samples. The simplicity of this BKS approach may not be detrimental, instead it has provided a fast and effective way of combining multiple predictions, as demonstrated in [29] for the classification of unconstrained handwritten numerals.

6 Summary

A promising approach that integrates the advantages of two neural network models into a unified framework has been described. The FAM network realizes a synthesis of ART and fuzzy set theory for tackling the stability-plasticity dilemma. However, it does not provide probability estimates for the target predictions. Hence the PNN is utilized to compensate for this limitation of FAM. In view of the suitability of the underlying learning methodologies and similarity of the network topologies, both FAM and the PNN have been combined to form a new hybrid network – PFAM – for on-line classification and prediction tasks. It alleviates drawbacks of both networks and, at the same time, maintains their self-organizing and incremental learning properties. Since the FAM learning methodology is adopted in PFAM, all properties of ART such as stable category formation and fast incremental learning are preserved. In the prediction phase, however, while FAM maintains the winner-take-all rule, the PFAM network applies all information coded in the network to estimate posterior probabilities of different target classes according to the Parzen-window probability estimation technique. Subsequently, statistical decision criteria such as the Bayes decision rule can be implemented in PFAM.

Issues on how to integrate both FAM and the PNN efficiently, along with some necessary modifications, have also been discussed. Incorporating procedures for estimating the center and width of the Gaussian kernels can further enhance the generalization ability of PFAM. In terms of computational efficiency, PFAM requires additional time to compute the centers and widths of the kernel functions, as well as storage to keep the additional center weight vectors. These demands are not present in FAM.

To further improve the robustness of the resulting system, a multiple classifier architecture comprising several PFAM networks can be formed. Then, outputs from a number of classifiers are combined using some decision combination procedure to yield an integrated result. Three decision combination algorithms, *i.e.*, the voting, Bayesian, and BKS methods are discussed for aggregating predictions from multiple PFAM classifiers. The efficacy of these three algorithms will be assessed using a number of real data sets and will be reported in Part II of our work.

Acknowledgments

The first author acknowledges the research grants provided by Universiti Sains Malaysia as well as the Ministry of Science, Technology, and the Environment Malaysia (No. 06-02-05-8002) that have in part resulted in this chapter.

References

[1] Rumelhart, D.E, Hinton, G.E., and Williams, R.J. (1986), "Learning internal representation by error propagation," in Rumelhart, D.E. and McLelland, J.L. (Eds.), *Parallel Distributed Processing, I*, pp. 318-362, Cambridge, MA: MIT Press.

[2] Broomhead, D.S. and Lowe, D. (1988), "Multivariate functional interpolation and adaptive networks," *Complex Systems*, Vol. 2, pp. 321-355.

[3] Cybenko, G. (1989), "Approximation by superposition of a sigmoidal function," *Mathematics of Control, Signals and Systems*, Vol. 2, pp. 303-314.

[4] Poggio, T. and Girosi, F. (1990), "Network approximation and learning," *Proceedings of IEEE*, Vol. 78, pp. 1481-1497.

[5] White, H. (1989), "Learning in artificial neural networks: a statistical perspective," *Neural Computation*, Vol. 1, pp. 425-464.

[6] Richard, M.D. and Lippmann, R.P. (1991), "Neural network classifiers estimate Bayesian *a posteriori* probabilities," *Neural Computation*, Vol. 3, pp. 461-483.

[7] Grossberg, S. (1980), "How does a brain build a cognitive code," *Psychological Review*, Vol. 87, pp. 1-51.

[8] Carpenter, G.A. and Grossberg, S. (1987), "A massively parallel architecture for a self-organizing neural pattern recognition machine," *Computer Vision, Graphics and Image Processing*, Vol. 37, pp. 54-115.

[9] Carpenter, G.A. and Grossberg, S. (1988), "The ART of adaptive pattern recognition by a self-organizing neural network," *IEEE Computer*, Vol. 21, pp. 77-88.

[10] Carpenter, G.A. and Grossberg, S. (1987), "ART 2: stable self-organization of pattern recognition codes for analogue input patterns," *Applied Optics*, Vol. 26, pp. 4919-4930.

[11] Carpenter, G.A. and Grossberg, S. (1990), "ART 3 hierarchical search: chemical transmitters in self-organizing pattern recognition architectures," *Neural Networks*, Vol. 3, pp. 129-152.

[12] Carpenter, G.A., Grossberg, S., and Rosen, D.B. (1991), "Fuzzy ART: fast stable learning and categorization of analogue patterns by an adaptive resonance system," *Neural Networks*, Vol. 4, pp. 759-771.

[13] Carpenter, G.A., Grossberg, S., and Reynolds, J.H. (1991), "ARTMAP: supervised real-time learning and classification of nonstationary data by a self-organizing neural network," *Neural Networks*, Vol. 4, pp. 565-588.

[14] Carpenter, G.A., Grossberg, S., Markuzon, N., Reynolds, J.H., and Rosen, D.B. (1992), "Fuzzy ARTMAP: a neural network architecture for incremental supervised learning of analogue multidimensional maps," *IEEE Transactions on Neural Networks*, Vol. 3, pp. 698-712.

[15] Carpenter, G.A. and Ross, W.D. (1995), "ART-EMAP: a neural network architecture for object recognition by evidence accumulation," *IEEE Transactions on Neural Networks*, Vol. 6, pp. 805-818.

[16] Specht, D.F. (1990), "Probabilistic neural networks," *Neural Networks*, Vol. 3, pp. 109-118.

[17] Zadeh, L. (1965), "Fuzzy sets," *Information and Control*, Vol. 8, pp. 338-353.

[18] Lim, C.P. and Harrison, R.F. (1997), "Modified Fuzzy ARTMAP approaches Bayes optimal classification rates: an empirical demonstration, *Neural Networks*, Vol. 10, pp. 755-774.

[19] Parzen, E. (1962), "On estimation of a probability density function and mode," *Annals of Mathematical Statistics*, Vol. 33, pp. 1065-1076.

[20] Cacoullos, T. (1966), "Estimation of a multivariate density," *Annals of the Institute of Statistical Mathematics (Tokyo)*, Vol. 18, pp. 179-189.

[21] Fukunaga, K. (1972), *Introduction to Statistical Pattern Recognition*, New York: Academic Press.

[22] Duda, R.O. and Hart, P.E. (1973), *Pattern Classification and Scene Analysis*, New York: John Wiley and Sons.

[23] Lim, C.P. and Harrison, R.F. (1997), "An incremental adaptive network for on-line supervised learning and probability estimation," *Neural Networks*, Vol. 10, pp. 925-939.

[24] Masters, T. (1993), *Practical Neural Network Recipes in C++*, San Diego, CA: Academic Press.

[25] Masters, T. (1995), *Advanced Algorithms for Neural Networks: A C++ Sourcebook*, New York: John Wiley and Sons.

[26] Musavi, M.T., Ahmed, W., Chan, K.H., Faris, K.B., and Hummels, D.M. (1992), "On the training of radial basis function classifiers" *Neural Networks*, Vol. 5, pp. 595-603.

[27] Xu, L., Krzyzak, A., and Suen, C.Y. (1992), "Methods of combining multiple classifiers and their applications to handwriting recognition," *IEEE Transactions on Systems, Man, and Cybernetics*, Vol. 22, pp. 418-435.

[28] Ho, T.K., Hull, J.J., and Srihari, S.N. (1994), "Decision combination in multiple classifier systems," *IEEE Transactions on Pattern Analysis and Machine Intelligence*, Vol. 16, pp. 66-75.

[29] Huang, Y.S., and Suen, C.Y. (1995), "A method of combining multiple experts for the recognition of unconstrained handwritten numerals," *IEEE Transactions on Pattern Analysis and Machine Intelligence*, Vol. 17, pp. 90-94.

[30] Pearl, J. (1988), *Probabilistic Reasoning in Intelligent Systems: Networks of Plausible Inference*, San Mateo, CA.: Morgan Kaufmann.

CHAPTER 7

ART-BASED AUTONOMOUS LEARNING SYSTEMS: PART II – APPLICATIONS

C.P. Lim
School of Industrial Technology
Universiti Sains Malaysia
11800 Penang
Malaysia

R.F. Harrison
Department of Automatic Control and Systems Engineering
The University of Sheffield
Mappin Street, Sheffield S1 3JD
United Kingdom

This chapter presents an evaluation of the practical applicability of the multiple classifier systems developed in Part I. First, a benchmark study is conducted to compare the system performance with other machine learning algorithms. Then, two medical decision-making problems comprising real patient records collected from a number of hospitals are investigated, *viz.* (i) the prognosis of patients exhibiting complications admitted to coronary care units, and (ii) the prediction of survival rates among trauma patients. Various operating strategies including off-line and on-line learning are examined, and implications of the results are discussed. In particular, the outcomes reveal the potential of employing such an autonomously learning system as a useful decision support tool in the medical arena.

1 Introduction

The availability of fast and yet economical computers has made the use of computer-aided systems increasingly attractive. In the clinical arena, such computerized systems have attracted much interest owing to their perceived potential to assist in and improve the decision-making process. The main thrust of our work is in the design and development of appropriate tools and methods for clinical diagnosis and prognosis. In particular, we concentrate on the use of artificial neural networks as an "intelligent" inference engine for the development of computerized decision aids. In addition to the obvious need for reliable and accurate decision making, a useful decision support system must also be robust, easy to use, and portable. Hence, the ability for on-line or causal learning is important so that the system can readily adapt to operate under various conditions, *e.g.*, in situations with markedly different demography, or changing clinical practice from one site to another. We, therefore, strive to design a system that is able to learn autonomously and continuously without human intervention with the ultimate goal of providing techniques which can allow domain experts, rather than computer experts, to develop their own intelligent decision support tools.

One of the difficulties in making an accurate medical prediction is the uncertainty associated with the information at hand. Frequently, many unconnected diseases may correspond to an identical set of data items comprising symptoms, measurements, and historical records. Conversely, any given diagnosis may correspond to a number of distinct data patterns. To tackle this problem, an intelligent decision support system can integrate data-based analytical techniques, such as information, decision, and classification theories, with modern knowledge-based methods to improve decision-making. Artificial neural networks possess both of these properties. Indeed, many medical problems have been attempted using neural-network-based systems, *e.g.*, the diagnosis of epilepsy [1], low back disorders [2], heart attack [3]-[5], breast cancer [6], [7], prostate cancer [8], and analysis of radiography images [9] and ECG data [10], [11].

In this chapter, the Probabilistic Fuzzy ARTMAP (PFAM) network, as presented in Part I, is applied to pattern classification problems. In particular, a number of PFAM-based classifiers are formed to improve the generalization and adaptability of PFAM. Before embarking on the medical experiments, a benchmark data set taken from a public domain repository is employed to evaluate the performance of the individual and multiple PFAM systems. The results are compared with those from a number of machine-learning algorithms as reported in the literature. Then, two real medical data sets are used to assess the applicability of the PFAM-based systems to (i) prognosis of patients exhibiting complications admitted to coronary care units, and (ii) predicting survival rates among trauma patients. Various operating strategies including off-line and on-line learning are also investigated to realize a robust decision support system. The results are quantified using some statistical tests, and compared with other methods commonly used in the medical arena.

2 A Benchmark Study

This study is concerned with the classification of Landsat Satellite images as used in the Statlog project [12]. The data set consists of a small sub-section (82×100 pixels) of a scene from the original satellite images, where each pixel covers an area of approximately 80×80 meters on the ground. One frame of the Landsat Multi-Spectral Scanner (MSS) imagery comprises intensities of four spectral bands of the same scene, in which two of the spectra are in the visible region (corresponding approximately to green and red regions) and two are in the (near) infra-red region.

This Landsat MSS data can be obtained from the UCI Repository of Machine Learning Databases and Domain Theories [13]. The data set is divided into two files: a training file of 4435 samples, and a test file of 2000 samples. Each data sample comprises 36 integer-valued attributes – 4 spectral values multiplied by 9 pixels in a 3×3 neighborhood. Given the multi-spectral values, the aim of the experiment is to predict the target output of the central pixel as belonging to one of the six classes – red soil, cotton crop, grey soil, damp grey soil, soil with vegetation stubble, and very damp grey soil [12].

The Landsat MSS data set serves as a challenging benchmark problem as it comprises real and noise-corrupted satellite images. In [14], this data set was used to assess the performance of Fuzzy ARTMAP (FAM) and Fusion ARTMAP – another supervised ART network for multi-channel data fusion and classification. Furthermore, many classification algorithms have been evaluated using this data set as part of the Statlog project. Thus, it is useful to compare the performance of the individual and multiple PFAM systems with other approaches.

2.1 Off-Line Learning

Before conducting the experiment, the training data were randomized to produce five differently ordered training sets. Then two experiments were conducted. The first one used five PFAM networks, each trained using a different set of training data, and the results were averaged across these five individual networks. In addition to the average results, the second experiment combined the outputs from the five individual networks using three decision combination procedures, *i.e.*, the voting, Bayesian, and Behavior-Knowledge Space (BKS) multiple classifier systems (MCSs) as presented in Part I. Two methods of representing input samples are investigated, *i.e.*,

(i) Concatenation – where the data samples are used in their original form. In this approach, multiple PFAM classifiers are first trained on different orderings of training samples, and then tested on the test samples. The predictions from these classifiers are combined using MCS to give an overall prediction.

(ii) Modularization – where each data sample is divided into groups of relevant attributes. In the training phase, each group of attributes of a training item is assigned to a PFAM network. In the test phase, the predictions from these classifiers, each on a group of attributes of a test item, are combined using MCS to give an overall prediction.

Unless otherwise stated, parameters associated with the PFAM network were set at their "default" values: the ART_a baseline vigilance parameter, $\overline{\rho_a} = 0.0$ (forced choice); learning rate, $\beta_a = 1.0$ (fast learning); choice parameter, $\alpha_a \approx 0.0$ (conservative mode); overlapping parameter, $r = 2.0$; and confidence threshold of MCSs, $\lambda = 0.0$.

2.1.1 Results of the Concatenation Approach

Table 1 lists the results of a variety of classification algorithms from the Statlog project [12], whereas Table 2 depicts the FAM results reported in [14] and the results from this experiment. The results of PFAM were averages of five independent runs. For the voting, Bayesian, and BKS MCSs, the five individual PFAM outputs were combined to produce an overall prediction. Two baseline vigilance values were tested, *i.e.*, a low value of $\overline{\rho_a} = 0.0$ for forming coarse prototype categories of input samples in ART$_a$ (F_2^a nodes), and a high value of $\overline{\rho_a} = 0.9$ for forming fine categories. As might be expected, a considerable improvement in accuracy was achieved with $\overline{\rho_a} = 0.9$. The trade-off lay in the increase in the number of ART$_a$ categories created by FAM and PFAM. The increase was about 8 fold and 6 fold, respectively, when the baseline vigilance was increased from $\overline{\rho_a} = 0.0$ to $\overline{\rho_a} = 0.9$. Nevertheless, the *k*-NN approach, which achieved the best accuracy (Table 1), had to store all 4435 training samples as prototype nodes [12]-[14]. Thus, comparatively, when operating at $\overline{\rho_a} = 0.9$, FAM and PFAM can achieve a slightly lower accuracy (1.6%) but with a higher degree of code compression.

In terms of classification accuracy, the three MCSs operating at $\overline{\rho_a} = 0.9$ outperform not only FAM and PFAM, but also all the algorithms listed in Table 1. The BKS method achieved the best results of 91.7% for $\overline{\rho_a} = 0.0$ and 94.5% for $\overline{\rho_a} = 0.9$. All the three MCSs were also able to improve the average results of individual PFAM networks.

With $\overline{\rho_a} = 0.0$, the performance of PFAM was relatively poor when compared with other approaches. The inferiority might be due to the small number of ART$_a$ categories (prototype patterns) being created in the system. Recall that PFAM uses the Parzen-window probability estimation procedure to predict a target output in the prediction phase. Accuracy of the estimated probability density function (pdf) depends on the number of prototypes available. The larger the number of prototypes, the more accurate the estimated pdf becomes. Theoretically, the pdf will converge asymptotically to the underlying function as the number of prototypes tends to infinity. As can be seen in Table 2, performance of PFAM improved considerably with $\overline{\rho_a} = 0.9$, in which case more than five hundred prototypes were created.

Table 1. Classification accuracy from various algorithms on the Landsat MSS data set [12, pp. 145]. The algorithms can be divided broadly into three categories: (i) statistical techniques – k-NN (k-nearest neighbors), ALLOC80, Quadisc (quadratic discriminants), SMART, Logdisc (logistic discriminants), Discrim (Linear discriminants), NaiveBay (Naive Bayes), CASTLE; (ii) decision trees and rule-based methods – CART (classification and regression tree), IndCART (Inverse CART), Baytree (Bayes tree), NewID, CN2, C4.5, Cal5, AC^2; (iii) neural networks – LVQ, DIPOL92, RBF, Back-prop, Kohonen, Cascade. The Default accuracy is calculated by categorizing all test samples to the class that has the highest number of samples, *i.e.*, the maximum *a priori* classification rule.

Algorithm	Accuracy (%)
k-NN	90.6
LVQ	89.5
DIPOL92	88.9
RBF	87.9
ALLOC80	86.8
CART	86.2
IndCART	86.2
Back-prop	86.1
Baytree	85.3
NewID, CN2, C4.5	85.0
Cal5	84.9
Quadisc	84.5
AC^2	84.3
SMART	84.1
Logdisc, Cascade	83.7
Discrim	82.9
Kohonen	82.1
CASTLE	80.6
NaiveBay	71.3
Default	23.1

2.1.2 Results of the Modular Approach

In this experiment, data items corresponding to each of the four spectral bands were extracted and assigned to one PFAM network. This resulted in four PFAM networks, each handling input attributes from one scanner. Table 3 shows the results, averaged over five runs, of the MCSs by combining outputs from the four individual PFAM networks.

Generally, the results are inferior to those of the concatenation approach as well as to those in Table 1. By examining the performance of each PFAM network, we notice that, despite poor results, there was a proliferation of prototype nodes. The worst results with the highest number of ART_a categories from a single network were 37.3% accuracy with 1063 nodes for $\overline{\rho_a} = 0.0$, and 42.2% accuracy with 1375 nodes for $\overline{\rho_a} = 0.9$, respectively. Nevertheless, the results improved with the use of MCSs. In particular, the BKS approach produced an increase of about 30% in accuracy from the individual network results for both $\overline{\rho_a} = 0.0$ and $\overline{\rho_a} = 0.9$.

Table 2. Performance of individual and multiple networks for the Landsat MSS data set. The FAM results are those reported in [14].

Algorithm		$\overline{\rho_a} = 0.0$	$\overline{\rho_a} = 0.9$
FAM	Accuracy (%)	83.0	89.0
	No. of ART_a Categories	89	704
PFAM	Accuracy (%)	81.4	89.0
	No. of ART_a Categories	87	518
Multiple Classifier System	Voting (%)	86.1	90.8
	Bayesian (%)	87.0	91.6
	BKS (%)	91.7	94.5

Table 3. Performance of multiple classifiers using the modular approach for the Landsat MSS data.

MCS	$\overline{\rho_a} = 0.0$	$\overline{\rho_a} = 0.9$
Voting	59.4%	65.6%
Bayesian	69.2%	74.8%
BKS	83.3%	86.1%

Deficiency of this modular approach is also experienced with Fusion ARTMAP where the best performance was about 70% accuracy, and the same problem of category proliferation was also observed [14]. Failure of the modular approach might be due to inter-spectral dependency in the satellite images. This Landsat data set has very large correlation between attributes of input samples [12]. By segmenting the attributes into different scanner bands and assigning them to one classifier, inter-spectral information is lost. For example, the "red" spectrum scanner might be insensitive to certain "green" frequencies, and would produce irregular images associated with different output classes when the region under inspection is mostly vegetation [14]. Without reinforcement information from the "green" spectrum scanner on vegetation imagery, input attributes to the "red" classifier would be more likely to mismatch the target classes, which in turn would result in more new categories being created to code the samples. This pitfall is avoided in the concatenation approach where data from all scanners are concatenated into a single input sample. The resulting input sample representation would be more consistent with the target classes, and therefore enables the network to achieve a better generalization with a reduced number of prototype nodes.

2.2 Dual-Mode Learning

One of the advantages of ART-based systems is the ability to learn on-line without the trouble of re-training. On-line learning imitates the condition of a human operating in a natural environment. A system with on-line learning starts with no information about the environment and gradually accumulates knowledge as more and more samples are presented. Each data sample is used as both a test item and a training item. The operation proceeds as follows. First an input sample is presented to ART_a, and a prediction will be produced at ART_b according to the prediction-phase algorithm. The prediction is compared with the target output at ART_b, thus yielding a classification result. Then learning ensues to associate the input sample with its target output according to the learning-phase algorithm.

In real applications, however, a series of trials on historical data is normally required before a system is allowed to become operational. As a result, we propose a *dual-mode* learning strategy that combines both the off-line and on-line learning procedures. Using this strategy, an off-line learning process is first applied to equip the system with a knowledge base before on-line learning is initiated. Here, each PFAM network is trained with a different ordering of the training set so that different prototypes can be established during the off-line learning phase. MCSs are implemented to integrate outputs from various networks for subsequent incoming samples during the on-line learning phase.

Since learning takes place incrementally in on-line learning mode, to reflect accumulation of knowledge through past experiences and to assess whether the performance improves through time, a 1000-sample moving window is applied to calculate the on-line results. For example, the accuracy at sample 2000 is the percentage of correct predictions from trials 1001-2000. Note that even though the problem under scrutiny may itself be stationary, the on-line predictions form a non-stationary process due to the build-up of templates (the so-called finite-operating-time problem).

Two sets of dual-mode learning experiments were conducted with $\overline{\rho_a}$ = 0.0 and $\overline{\rho_a}$ = 0.9. As explained earlier, five independent PFAM classifiers were initially trained with 1000 samples, each with a different ordering of data. These classifiers were then tested, on-line, on the remaining 5435 samples, and their results were combined using MCS.

Figure 1 shows the overall classification accuracy against increasing number of input samples for the average results of five independent classifiers, as well as the voting, Bayesian, and BKS MCSs. Some observations can be drawn from the results: (i) performance improves with a high vigilance value; (ii) all the three MCSs perform better than the individual classifiers; (iii) the BKS approach produces the best performance, followed by the Bayesian approach and then the voting method. Nevertheless, with $\overline{\rho_a}$ = 0.9, both the Bayesian and voting methods achieved virtually the same performance.

In comparison with the off-line results, this dual-mode learning strategy achieved a slightly inferior performance. However, the system has the advantage of learning safely on-line, and the learning process is on going in possibly non-stationary environments. The problems associated with off-line learning such as pre-determined network size and re-training are thus avoided. Besides, all the classifiers do not have to start from a naive condition as in purely on-line operations. Instead, they are equipped with some "knowledge" about the environment before they are put into use. In addition, decisions from multiple classifiers can be combined to produce a more reliable classification system. With the ability of incremental learning, the system can improve its performance autonomously when it encounters more and more incoming samples over time.

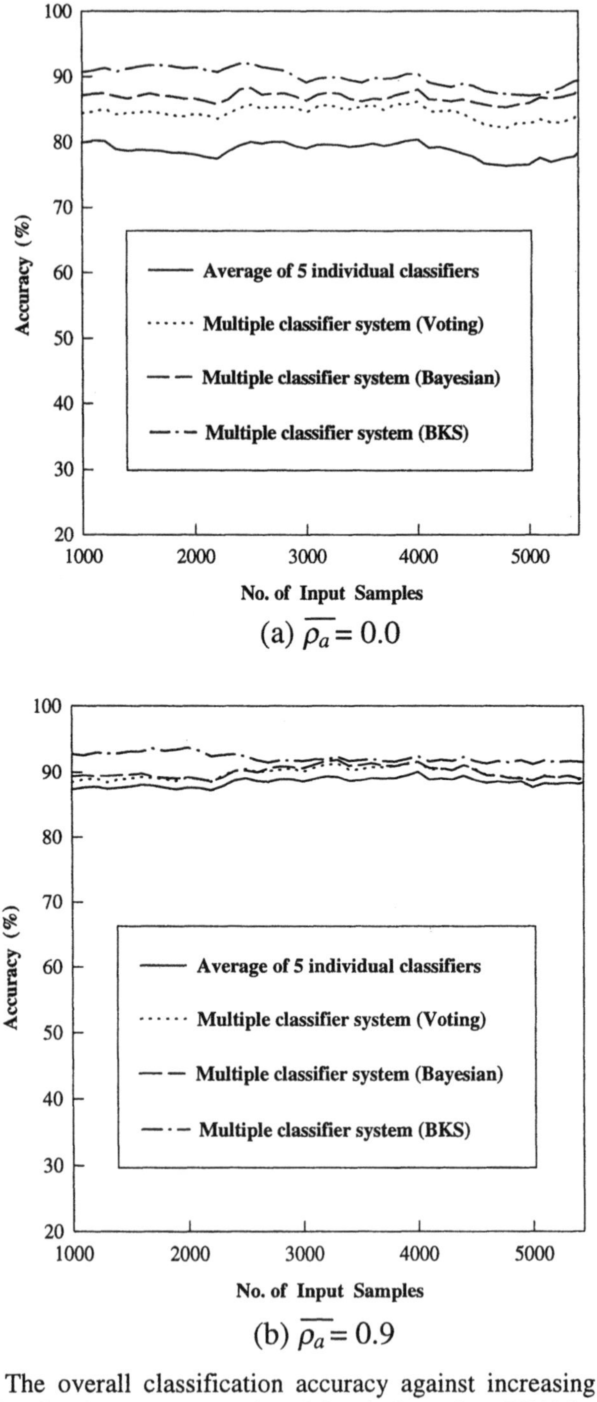

Figure 1. The overall classification accuracy against increasing number of input samples for the average results of five independent PFAM classifiers as well as the voting, Bayesian and BKS multiple classifier systems.

3 Medical Prognosis

Here, two medical databases involving real patient records were employed to demonstrate the applicability of the individual and multiple PFAM networks to medical decision support tasks. All the experiments described below employed the same parameter settings as in the benchmark study, and input concatenation was used. To quantify the results, three performance measures that are commonly used in medical diagnostic systems are calculated, *i.e.*,

- Accuracy (ACC) – the ratio of the number of correct diagnoses to the total number of cases;

- Sensitivity (SENS) – the ratio of the number of correct positive diagnoses to the total number of patients having a positive outcome;

- Specificity (SPEC) – the ratio of the number of correct negative diagnoses to the total number of patients having a negative outcome.

3.1 Prognosis of Complications

Since the 1950s, there has been a progressive reduction in the recommended length of hospital stay for patients in a coronary care unit (CCU), which has provided savings in terms of resources and costs without significantly increasing mortality rates [15]. However, early hospital discharge requires exact identification of patients who are subject to minimal risk of death. Thus, an accurate system for prognosis of patients in CCU might improve the planning and management of hospital facilities as well as healthcare of patients. The application highlighted here is the prediction of complications in CCU patients at the Leicester Royal Infirmary, United Kingdom. The study included a total number of 3721 patients admitted between August 1988 and August 1992. Information such as symptoms, measurements, and risk factors were recorded in a proforma for each patient. Nine clinical and electrocardiographic data items that are significant in the prognosis were extracted and analyzed in this study. The final outcome, assigned by senior clinicians concerned with the patient's care, is either one of the following:

(i) Class C_1 – patient is at risk of serious complications, or of high Creatine Kinase assay (> 1000 U / 1) indicative of myocardial infarction (heart attack), or death

(ii) Class C_2 – none of the above risks

3.1.1 Off-Line Learning

The database was divided into a training set of 1000 samples and a test set of 2721 samples. An ensemble of twenty-five PFAM networks was tested, and a total number of twelve MCSs was formed, each with five networks selected randomly from the pool of individual networks and with at least four different ones in each case. This data set has been analyzed previously using a logistic regression model [16]. The results were assessed with the Receiver Operating Characteristic (ROC) curve – a widely used tool for evaluating performance of medical diagnostic systems [17]-[20]. The area under the ROC curve achieved by the logistic regression model was 79.0%.

Table 4. Off-line learning results for the database of CCU patients. Figures in parentheses indicate the maximum and minimum results.

Algorithm		ACC	SENS	SPEC
Logistic regression (%)		73.0	73.1	72.9
PFAM (%)		69.3 (64.2-72.5)	69.2 (64.6-73.0)	69.3 (63.4-73.3)
Multiple Classifier System	Voting (%)	71.5 (70.2-73.6)	71.1 (68.4-78.3)	72.2 (65.1-74.6)
	Bayesian (%)	72.0 (69.2-74.1)	86.4 (82.2-91.7)	45.6 (31.8-52.0)
	BKS (%)	74.1 (73.3-74.6)	79.3 (74.6-84.4)	64.4 (55.2-71.0)

Table 4 summarizes all the results in this experiment in terms of accuracy, sensitivity, and specificity. The ROC area formulated by PFAM was 75%. Although a single run from PFAM could achieve a result equivalent to that of logistic regression, on average, the PFAM results were about 4% inferior compared to those of the logistic regression model. When experimenting with the Multi-layer Perceptron and Radial Basis Function networks, we also failed to attain a significantly better performance than the logistic regression model. Thus, the logistic regression model might have performed optimally for this data set.

Note that the probabilistic estimates of target outputs from individual PFAM networks was regulated using ROC analysis so that an approximately balanced score can be obtained for ACC, SENS, and SPEC [17]. This manipulation is not feasible in MCSs because their outputs are not expressed in terms of probabilities. Nonetheless, the voting strategy could still maintain a close level for the three indices, as shown in Table 4. The same observation had been encountered in another medical data set (the early diagnosis of heart attack) using FAM with voting [4]. The loss of the leveling effect in Bayesian and BKS MCSs is expected as these two algorithms only attempt to optimize the overall accuracy, *i.e.*, by the use of confusion matrices and focal units, respectively.

3.1.2 Dual-Mode Learning

For the dual-mode learning experiment, first twenty-five individual PFAM networks were trained with 1000 samples in random order, and then learning continued on-line for the remaining 2721 samples. Twelve MCSs, each comprising five networks with at least four different ones, were constructed. As pointed out earlier, outputs from the MCSs cannot be manipulated using the ROC to achieve a balanced score for ACC, SENS, and SPEC. Thus, we compare the performance based only on the accuracy (calculated using a 1000-sample moving window), as shown in Figure 2.

To quantify the on-line results statistically, estimates of the 95% confidence interval using the *t*-distribution for small sample analysis [21] are computed and plotted as error bars. The results of the three MCSs were significantly better than the average results of individual networks with clear-cut delineation of confidence intervals. With the help of the confidence bounds, one can see that the BKS approach exhibits, statistically, the best performance, followed by the Bayesian and then voting methods. However, the confidence intervals between the Bayesian and voting MCSs overlap at times. Once again, these results suggest, empirically, that multiple classifiers are able to outperform individual classifiers.

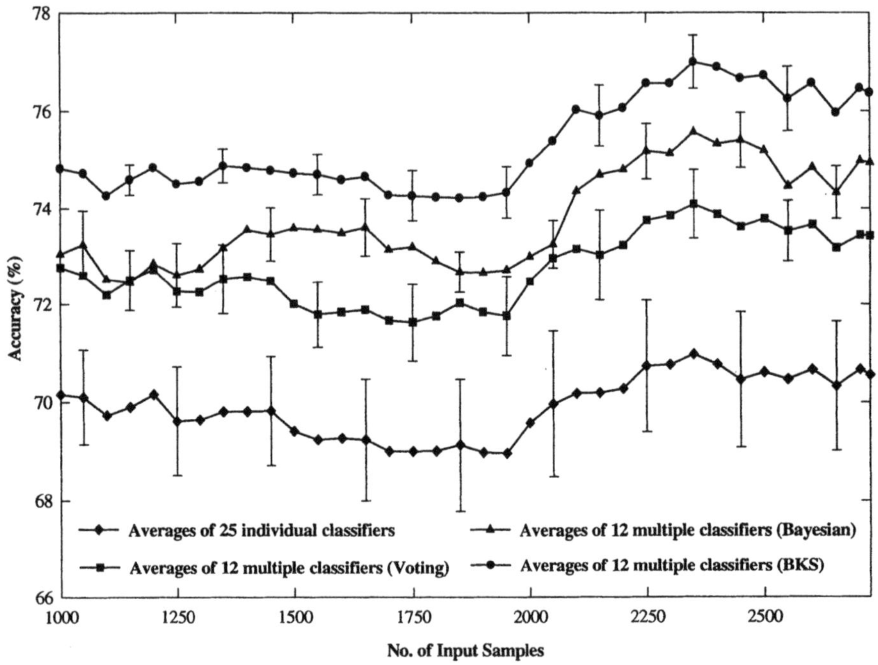

Figure 2. A comparison of accuracy of individual and multiple PFAM systems for the database of CCU patients. The error bars indicate estimates of their respective 95% confidence intervals.

3.2 Prognosis of Survival Rates

There is a widely held view that the standard of healthcare of trauma patients in the United Kingdom is sub-optimal. This belief is supported by evidence from preliminary reports of the UK Major Trauma Outcome Study (MTOS) [22], and the Scottish Trauma Audit Group [23]. Currently the most commonly used statistical technique for auditing trauma care is the TRISS method [24]. This methodology combines anatomical and physiological information in the administration of trauma diagnosis, which includes Injury Severity Scoring (ISS), Revised Trauma Score (RTS), patient age, and two injury types – blunt or penetrating [22], [24]. The patient's age is assigned a value of zero if the patient is under 55 years old and a value of one if he/she is 55 years old or over. The probability of survival is then estimated using logistic regression under the TRISS methodology.

In this study, a database of 6321 trauma patient records was employed. The patient records were collected between February 1993 and February 1995 from six Scottish teaching hospitals – Aberdeen Royal Infirmary, the Royal Infirmary of Edinburgh, Glasgow Royal Infirmary, the Western Infirmary of Glasgow, the Victoria Infirmary of Glasgow, and the Southern General Hospital of Glasgow. The aim of this study is to examine the prediction of individual and multiple PFAM systems in categorizing trauma patients into two classes:

(i) Class C_1 – survivors (ii) Class C_2 – non-survivors

Distribution of this database is highly skewed with only about 5% of C_2 instances. This deficiency of instances of one class compared with the other means that a classifier is likely to bias prediction towards the commonly seen class (C_1). Since approximately 95% of patients are in class C_1, the baseline performance of a classifier would be 95% accuracy, *i.e.*, by categorizing all patients as C_1 using the maximum *a priori* rule. Owing to the skewness of the data distribution, it is difficult to manipulate the probabilistic estimates of PFAM to maintain a close score for ACC, SENS, and SPEC. As a result, it was decided to operate PFAM at its "default" ROC curve operating threshold of 50% in the experiments described below.

3.2.1 Off-Line Learning

The database was divided into a training set of 2000 samples, and a test set of 4321 samples. Each data sample consisted of the four attributes used in TRISS – ISS, RTS, age, and type of injury. First, an experiment using the TRISS methodology was conducted using the coefficients of logistic regression provided by MTOS [22]. Then the same setup as in the CCU patient experiment was carried out. The results in terms of accuracy, sensitivity, and specificity are summarized in Table 5. The areas under the ROC curve achieved by TRISS and PFAM were 94.1% and 94.2% respectively.

Table 5. Off-line learning results for the database of trauma patients. Figures in parentheses indicate the maximum and minimum results.

Algorithm		ACC	SENS	SPEC
TRISS (%)		96.6	99.0	55.3
PFAM (%)		96.3 (93.5-97.4)	98.0 (94.7-99.7)	61.0 (46.5-73.0)
Multiple Classifier System	Voting (%)	97.4 (97.1-97.5)	99.1 (98.7-99.5)	62.3 (58.0-65.5)
	Bayesian (%)	97.1 (96.7-97.3)	99.7 (99.6-99.9)	42.5 (33.0-49.0)
	BKS (%)	97.6 (97.5-97.7)	99.3 (99.1-99.6)	62.5 (59.0-66.5)

Unlike the experiments on the CCU patients, PFAM was able to achieve comparable results with those of TRISS in terms of accuracy, sensitivity, specificity, as well as area under the ROC curve. As shown in Table 5, MCSs successfully improved the results from individual networks. Notice that the voting and BKS MCSs were able to improve all three performance indices. However, the Bayesian MCS achieved relatively high results for sensitivity and low results for specificity. As suggested in [25], stability of the Bayesian approach relies on the confusion matrices. If the confusion matrices of individual classifiers are well learned, then it can produce good results. By inspecting the confusion matrices, it was found that the elements were biased towards the *survival* class (as might be expected) which, in turn, caused the trade-off between sensitivity and specificity.

3.2.2 Dual-Mode Learning

The dual-mode learning strategy has been shown to be able to achieve good performance in previous experiments. The same setup as before was again used in this experiment. A set of 2000 samples was used for training, and the subsequent 4321 samples were applied for on-line learning.

Figure 3 shows a comparison of classification accuracy among the three MCSs and the individual networks. The results are bounded by their respective 95% confidence intervals. The BKS approach demonstrated a superior performance among the three decision combination algorithms. Performance of the voting MCS is similar to that of the Bayesian MCS. This might be owing to the poorly learned confusion matrices of the Bayesian decision combination method. Nevertheless, all MCSs performed better than individual classifiers with clear-cut delineation of confidence bounds, thus ascertaining again the benefit of using multiple classifiers in realizing a high-performance classification system.

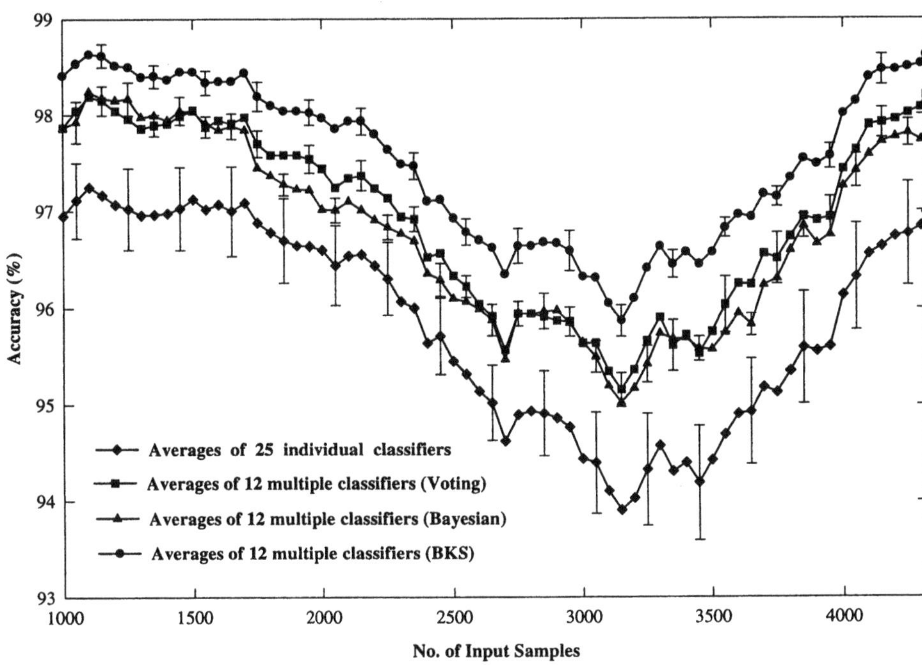

Figure 3. A comparison of accuracy between individual and multiple PFAM systems for the database of trauma patients. The error bars indicate estimates of their respective 95% confidence intervals.

4 Summary

A composition of individual and multiple classifiers formed using the PFAM network has been studied for pattern classification problems. The MCS aggregates outputs from disparate classifiers using some decision combination algorithm to yield a combined result. Three algorithms, *i.e.*, the voting, Bayesian, and BKS methods, have been tested, and their efficacy was examined using a benchmark database and two real medical databases collected from a number of hospitals in the United Kingdom. In general, it is found that multiple classifiers are able to enhance the performance of individual classifiers in all the experiments.

Good performance was reported for individual as well as multiple PFAM networks in the Landsat MSS benchmark study. Apart from the usual concatenation approach of grouping all input attributes into a high-dimensional vector, a modular approach that segregates the input vector into groups of related attributes has been tested. Predictions from different groups of attributes are combined through MCSs. However, the use of this modular approach is strongly dependent on the correlation among input attributes. If strong correlation exists, the modular approach not only will diminish the overall performance but will also induce unnecessary complexity in the PFAM network.

Since PFAM is able to learn incrementally, one practical strategy that comprises a mixture of off-line and on-line learning has been investigated. This dual-mode learning approach first allows a pool of PFAM networks to be trained with a set of training samples with arbitrary order to establish a knowledge base in each network. Then on-line learning is engaged in which predictions from multiple networks are combined using some decision combination algorithm. Therefore, a robust, adaptive incrementally learning classifier system can be realized to overcome problems associated with off-line learning networks, such as the need to pre-specify a static network size and re-training in non-stationary environments.

The medical prognosis experiments utilize databases of CCU and trauma patients – prime examples which potentially could benefit from a computerized decision support system to assist in the healthcare of patients. The results were compared to those from the logistic regression model and TRISS. Furthermore, confidence intervals were computed to quantify the performance statistically in dual-mode learning. The BKS approach repeatedly exhibited the best performance, whereas voting showed consistent results throughout the experiments. However, performance of the Bayesian method can be sensitive depending on how the confusion matrices have been constructed, as demonstrated in the experiments using the database of trauma patients.

From the experiments, the potential of dynamic neural networks, such as PFAM, as building blocks for devising systems that are able to learn incrementally and autonomously is evident. However, this is only the beginning of the development of genuinely intelligent systems to achieve the ultimate goal of allowing domain experts, such as doctors and clinicians, to design decision support tools according to their own requirements. Here, a pragmatic approach has been adopted to evaluate and confirm the applicability of the PFAM-based systems in various environments. Indeed, the PFAM network and its associated MCSs would benefit from a formal theoretical analysis of their dynamics. This would be a challenging task because they are non-linear on-line learning systems operating on random inputs in non-stationary environments. In addition, some practical aspects also need further investigation, *e.g.*, how to extract rules from PFAM for it to "explain" its predictions, and the use of other decision combination algorithms to improve the efficacy of multiple classifier systems.

Acknowledgments

The first author acknowledges the research grants provided by Universiti Sains Malaysia as well as the Ministry of Science, Technology, and the Environment Malaysia (No. 06-02-05-8002) that have in part resulted in this chapter. Thanks are due to Lee Kennedy and Kent Woods for the medical data and help in its interpretation.

References

[1] Apolloni, B., Avanzini, G., Cesa-Bianci, N., and Ronchini, G. (1990), "Diagnosis of epilepsy via backpropagation," *Proceedings of the International Joint Conference on Neural Networks*, Vol. 2, pp. 571-574.

[2] Bounds, D., Lloyd, P., and Mathew, B. (1990), "A comparison of neural network and other pattern recognition approaches to the diagnosis of low back disorders," *Neural Networks*, vol. 3, pp. 583-591.

[3] Harrison, R.F., Marshall, S.J., and Kennedy, R.L. (1991), "A connectionist approach to the early diagnosis of myocardial infarction," *Proceedings of the Third European Conference on Artificial Intelligence in Medicine (AIME-91)*, pp. 119-128.

[4] Harrison, R.F., Lim, C.P., and Kennedy, R.L. (1994), "Autonomously learning neural networks for clinical decision support," in Ifeachor, E.C. and Rosen, K.G. (Eds.), *Proceedings of the International Conference on Neural Networks and Expert Systems in Medicine and Healthcare (NNESMED-94)*, pp. 15-22, Plymouth: University of Plymouth.

[5] Baxt, W.G., (1990), "Use of an artificial neural network for data analysis in clinical decision support: the diagnosis of acute coronary occlusion," *Neural Computation*, Vol. 2, pp. 480-489.

[6] Ravdin, P.M. and Clark, G.M. (1992), "A practical application of neural network analysis for predicting outcome of individual breast cancer patients," *Breast Cancer Research Treat*, Vol. 22, pp. 285-293.

[7] Downs, J., Harrison, R.F., Kennedy, R.L., and Cross, S.S. (1996), "Application of the Fuzzy ARTMAP neural network model to medical pattern classification tasks," *Artificial Intelligence in Medicine*, Vol. 8, pp. 403-428.

[8] Snow, P.B., Smith, D.S., and Catalona, W.J., (1994), "Artificial neural networks in the diagnosis and prognosis of prostate cancer: a pilot study," *Journal of Urology*, Vol. 152 (5 Pt 2), pp. 1923-26.

[9] Lin J.S., et al., (1993), "Application of neural networks for reduction of false-positive detections in digital chest radiographs," *Proceedings of Annual Symposium on Computers Applied to Medical Care*, pp. 434-438.

[10] Edenbrandt, L., Helen, B., and Fahlm, O. (1993), "Neural networks for analysis of ECG complexes," *Journal of Electrocardiology*, Vol. 26, pp. 74.

[11] Bortolan, G. and Willems, J.L. (1993), "Diagnostic ECG classification based on neural networks," *Journal of Electrocardiology*, Vol. 26, pp. 75-79.

[12] Michie, D., Spiegelhalter, D.J., and Taylor, C.C. (1994), *Machine Learning, Neural and Statistical Classification*, Oxford: Oxford Press.

[13] Blake, C., Keogh, E., and Merz, C.J. (1998), *UCI Repository of machine learning databases* [http://www.ics.uci.edu/~mlearn/MLRepository.html]. Irvine, CA: University of California, Department of Information and Computer Science.

[14] Asfour, R.Y. (1995), *Fusion ARTMAP: neural networks for multisensor fusion and classification*, Ph.D. Thesis, Boston University.

[15] Parsons, R.W., Jamrozik, K.D., Hobbs, M.S.T. and Thompson, D.L. (1994), "Early identification of patients at low risk of death after myocardial infarction and potentially suitable for early hospital discharge," *British Medical Journal*, Vol. 308, pp. 1006-1010.

[16] Kennedy, R.L., Harrison, R.F., Fraser, H.S., Fletcher, S., Burton, A.M., McStay, L.N., and Woods, K.L. (1996), *Prediction of severe complications and mortality in patients admitted to a coronary care unit*, Research Report No. 655, Department of Automatic Control and Systems Engineering, University of Sheffield.

[17] Swets, J.A. and Pickett, R.M. (1982), *Evaluation of Diagnostic Systems*, London: Academic Press.

[18] Hanley, J.A. and McNeil, B.J. (1982), "The meaning and use of the area under a Receiver Operating Characteristic (ROC) Curve," *Radiology*, Vol. 143, pp. 29-36.

[19] Hanley, J.A. and McNeil, B.J. (1983), "A method for comparing the areas under Receiver Operating Characteristic curves derived from the same cases," *Radiology*, Vol. 148, pp. 839-843.

[20] Meistrell, M. (1990), "Evaluation of neural network performance by Receiver Operating Characteristic analysis: examples from the biotechnology domain," *Computer Methods and Programs in Biomedicine*, Vol. 32, pp. 73-80.

[21] Bland, M. (1987), *An Introduction to Medical Statistics*, Oxford: Oxford University Press.

[22] Yates, D.W., Woodford, M., and Hollis, S. (1992), "Preliminary analysis of the care of injured patients in 33 British hospitals: first report of the United Kingdom major trauma outcome study," *British Medical Journal*, Vol. 305, pp. 737-740.

[23] Little, K. *et al.* (1995), "Preliminary analysis of the care of injured patients in five Scottish teaching hospitals: first report from the Scottish Trauma Audit Group (STAG)," *Health Bulletin*, Vol. 53, pp. 55-65.

[24] Boyd, C.R., Tolson, M.A., and Copes, W.S. (1987), "Evaluating trauma care: the TRISS method," *Journal of Trauma*, Vol. 27, pp. 370-378.

[25] Xu, L., Krzyzak, A., and Suen, C.Y. (1992), "Methods of combining multiple classifiers and their applications to handwriting recognition," *IEEE Transactions on Systems, Man, and Cybernetics*, Vol. 22, pp. 418-435.

CHAPTER 8

ADAPTIVE RESONANCE THEORY MICROCHIPS

T. Serrano-Gotarredona and **B. Linares-Barranco**
National Microelectronics Center, Sevilla
Spain
{terese,bernabe}@imse.cnm.es

This chapter addresses implementation of learning models inspired by Adaptive Resonance Theory. A real-time clustering microchip based on the ART1 algorithm is presented, capable of classification and fast learning of 100-bit input patterns into up to 18 categories. A second chip is also presented which implements the ARTMAP algorithm for pattern association. We include experimental learning results from both prototypes, fabricated in standard single-poly double-metal $1.6\mu m$ and $1.0\mu m$ CMOS digital processes. MOS Transistors mismatch characterization issues are also addressed.

1 Introduction

Two types of neural hardware engineering can be distinguished. The first involves designing "general purpose" hardware accelerators or systems that speed up neural algorithms running on conventional computers [1], [2], [3]. This kind of hardware allows considerable flexibility in the topology and operations of the neural systems. In this way algorithm researchers have a powerful tool to further develop neural algorithms and industry engineers have some attractive chips that significantly speed up their neural commercial products. The second involves designing a real-time system for a specific application. They must select the best-suited algorithm and map it into hardware. This achieves a close-to-optimum efficient hardware for a limited range of applications. The work described in this chapter falls into this second category of hardware engineering. The specific application is real-time clustering of input patterns.

A clustering device is able to build categories from a collection of patterns. A real-time clustering device has to be able to do this at the speed of arrival of the patterns. There are some clustering algorithms [4], [5], [6], [7], [8], [9] that use off-line training to build the categories. For a real-time clustering device, however, it would be desirable to use an algorithm that can be trained on-line: if a new pattern arrives, the algorithm updates its internal knowledge (instead of erasing all the accumulated knowledge and retrain with the old and new collection of patterns).

For the second type of neural hardware engineers, the issue of efficiently implementing in hardware a real size neural network is not a trivial task. Many neural network algorithms are available in the literature which have been developed, studied, and optimized for applications through computer and/or software based systems. Consequently, when designing a hardware realization, engineers face many problems like excessive interconnectivity, high resolution of weights, high precision of operations, complicated operator requirements (e.g., integrals and derivatives), and high number of neurons required for a real-world application. Many times some of these requirements can be relaxed, the topology modified, or the operations simplified, with no significant deterioration of global operation of the neural system but with a considerable boost in the hardware performance. Modifying neural algorithms to make them more VLSI-friendly and produce more efficient hardware should be a common practice among neural hardware engineers of the second type [10], [11]. After selecting an appropriate neural algorithm, the next step consists of studying how far the algorithm can be simplified without performance degradation. The simplifications have to be hardware-oriented, so that the final combination of "theoretical algorithm" + "hardware circuit technique" results in a high performance real time system. The success of the hardware system depends on the selection of the algorithm, the selection of a powerful circuit design technique, and how the algorithm is modified to efficiently "marry" the circuit technique resulting in an optimum performance final system.

In this chapter, Adaptive Resonance Theory algorithms are considered mainly due to their attractive hardware-oriented properties, as well as the theoretical computational properties. We will concentrate on the implementation of ART1 based systems (i.e., ART1 and ARTMAP) which work on binary valued input patterns. As we will describe later, we also chose to slightly modify the mathematical algorithm to obtain more efficient hardware. This modification allows the use of simpler operations while preserving all the computational properties of the original architecture. As an extra bonus, the hardware circuit introduces a significant speed improvement as it automatically parallels the sequential ART search process (checking of the vigilance criterion). The advantageous computational and hardware oriented features of the ART1 algorithm are described in the following two sections, respectively.

2 Computational Properties of the ART1 Algorithm

From a purely algorithmic point of view, the ART1 architecture is capable of learning the recognition codes, in an unsupervised way, in response to arbitrary orderings of arbitrarily many and complex binary input patterns. This architecture has a collection of interesting computational properties [12], [13]:

- **Self-Scaling:** The self-scaling property discovers critical features in a context-sensitive way.
- **Vigilance or Variable Coarseness:** There is a vigilance parameter ($0 \leq \rho < 1$) that allows tuning the coarseness of the categories to learn.
- **Subset and Superset Direct Access:** The system is able to classify a new input pattern as belonging to either a subset or a superset category, depending on global similarity criteria. No restrictions on input orthogonality or linear predictability are needed.
- **Stable Category Learning:** In response to an arbitrary list (finite or infinite) of binary input patterns, learning is assured to self-stabilize within a finite number of learning trials.
- **Biasing the Network to Form New Categories:** There is a parameter that can bias the tendency of the system to code unfamiliar patterns into new categories, independent of the vigilance parameter.

- **On-Line Learning:** The ART1 algorithm learns as it performs, as opposed to other algorithms, where first the algorithm must be trained and second, it can be used in an application. The ART1 algorithm can incorporate new knowledge as it is being used. This property makes ART1 an excellent candidate for real-time clustering.

- **Capturing Rare Events:** ART1 is able to identify and build clusters of events that appear with a very low frequency. Even if an event corresponding to a clearly distinct cluster appears only once, ART1 is able to detect it while building and preserving the corresponding cluster or category.

3 Hardware-Oriented Attractive Properties of the ART1 Algorithm

3.1 Fast Learning

In performance comparison of hardware implementations, a common figure of merit is the number of interconnections per second. More refined figures have been proposed that include resolution and precision [14]. However, these figures would be reasonably fair criteria for the first type of hardware engineering mentioned above, the general-purpose one. In order to compare hardware systems of the second type, the specific-application neural hardware, some global figure must be used that evaluates the overall system performance. Usually this figure will be application dependent. In this Chapter, since we are concerned with real-time clustering of binary input patterns, an appropriate figure of merit might be

$$ppc/s = \frac{\text{number of patterns processed}}{seconds} \times pixels \times categories \qquad (1)$$

where,

- **number of patterns processed/second** is the speed at which patterns are classified and learned (including the number of learning trials required). This speed generally depends on the patterns themselves, and on the knowledge already stored in the system. Therefore, this speed can be given as an average or as the slowest case measured.

- **pixels** is the maximum number of pixels of the input patterns.
- **categories** is the maximum number of categories the system is able to form.

As we will see later in "Experimental Results", one of the chips is able to cluster up to 18 different categories of binary patterns with 100 pixels, while classifying and learning each pattern in less than $1.8\mu s$. Since ART1 learns on-line, 1 iteration of input patterns presentations provides the system with sufficient knowledge to perform properly. This results in a *ppc/s* of

$$\begin{aligned} ppc/s = & \frac{n \text{ patterns}}{1 \text{ iteration} \times n \text{ patterns} \times 1.8\mu s} \times 100 \text{ pixels} \times 18 \text{ categories} = \\ & = 1.0 \times 10^9 ppc/s \end{aligned} \quad (2)$$

If we would like to obtain the same performance using Backpropagation based hardware, and assuming the network would learn with 10,000 iterations of patterns presentations, this means that a speed of $180ps$ would be needed for each pattern classification and corresponding weights update. Assuming this task could be performed with a Backpropagation network with 100 input neurons, 5 hidden-layer neurons, and 5 output neurons[1] (which means a total of $(100 \times 5) + (5 \times 5) = 525$ interconnections), and that the speed of feedforward classification is the same as for feedback learning, hardware able to perform

$$\frac{2 \times 525 \text{ connections}}{180ps} = 5.83 \times 10^{12} \text{ connections/s + connection-updates/s} \quad (3)$$

would be needed. For the chips described in this Chapter, since they are based on the powerful ART1 algorithm, the above performance can be achieved with a hardware of only 4.4×10^9 *connections/s* plus *connections-updates/s*.

Note that the Backpropagation algorithm is not appropriate for clustering applications, and comparing it against ART1 is slightly unfair. There are other algorithms available in the literature that have been developed

[1]Optimistically, a Backpropagation network with 5 output neurons could code up to 32 categories

specially for clustering applications. However, they usually do not provide all the computational properties mentioned in the previous section, specially the "*On-Line Learning*" property which is crucial for real-time clustering, or they present serious difficulties when trying to map them into hardware.

3.2 Binary Memory

Another hardware attractive feature that an ART1 based implementation offers with respect to others, is that the interconnection weights do not have to be analog. Most of the neural algorithms reported in the literature require a real-valued set of weights defined within a certain interval. These weights can be discretized in a number of digital steps, but the granularity required for proper operation of the system is usually very fine (around 16-bits for the Back-Propagation algorithm [15]). Even worse, in some cases the granularity requirements become more severe as the size of the system increases. For example, in a BAM system [16] of $N \times M$ neurons, storage capacity has been heuristically estimated to be around $n_p = (N \times M)^{1/4}$ [17], where n_p is the average maximum number of patterns that can be stored. The resolution required by the interconnection weights in this case is at least $n_p + 1$. For chips based on the ART1 algorithm, since they require only binary-valued weights, the resolution of the weights is not affected by the size nor the storage capacity of the system. This, and the non necessity of analog weights is one of the most hardware attractive features of the ART1 algorithm.

3.3 Scaling

Another consideration to take into account during the design of a hardware system is how it scales up with size and performance. We have already mentioned that some neural systems need to increase their weight resolution as they scale up. Another feature is how their size and interconnectivity scale up with pattern size or storage capacity. For an ART1 based system, the number of neurons N in the bottom layer is the number of pixels of the patterns, the number of neurons M in the top layer is the maximum number of categories, and $N \times M$ is the number of synapses. Therefore, this system scales up linearly with storage capacity (M) and input pixels (N). For a BAM system, for example, the size scales quadratically with the storage capacity and the number of pixels [17].

In the case of an analog hardware, random and systematic errors due to fabrication process variations will appear. A neural network can usually cope very well with random errors, even if the size of the system increases. However, systematic errors may accumulate as the system increases and may render the complete network useless as it scales up. For example, if using transconductors based circuits (T-mode) [11], [18], due to the limited input voltage range of the transconductors (around $100mV$ if they have to be of minimum area) the practical maximum number of synapses is very limited (if each synapse introduces $1mV$ of systematic offset, 50 synapses would completely saturate the transconductor but misfunctions would appear with less synapses). The chosen circuit technique must be either insensitive to the accumulation of systematic errors, or allow for some kind of calibration technique to overcome them.

4 Description Levels of the ART1 Algorithm

In the original ART1 paper [12], the architecture is mathematically described by sets of Short Term Memory (STM) and of Long Term Memory (LTM) time domain nonlinear differential equations. A valid assumption made by Carpenter and Grossberg, is to make the STM differential equations settle instantaneously to their corresponding steady state, and consider only the dynamics of the LTM differential equations. In this case, the STM differential equations must be substituted by nonlinear algebraic equations that describe the corresponding steady state of the system. Furthermore, Carpenter and Grossberg also introduced the fast learning mode of the ART1 architecture, in which the LTM differential equations are also substituted by their corresponding steady-state nonlinear algebraic equations. Thus the ART1 architecture originally modelled as a dynamically evolving collection of neurons and synapses governed by time-domain nonlinear differential equations, can be behaviorally modelled as the sequential application of nonlinear algebraic equations: an input pattern is given, the corresponding STM steady state is computed through the STM algebraic equations, and the system weights are updated using the corresponding LTM algebraic equations.

At this point three different levels of ART1 implementations (both in software or in hardware) can be distinguished:

- **Type-1:** Full Model Implementation: Both STM and LTM time-domain differential equations are realized. This implementation is the most expensive (both in software and in hardware), and requires a large amount of computational power.
- **Type-2:** STM Steady-State Implementation: Only the LTM time-domain differential equations are implemented. The STM behavior is governed by nonlinear algebraic equations. This implementation requires less resources than the previous one. However, proper sequencing of STM events must be introduced artificially, which is architecturally implicit in the Type-1 implementation.
- **Type-3:** Fast Learning Implementation: This implementation is computationally the least expensive. In this case, STM and LTM events must be artificially sequenced.

Regarding hardware implementations of the ART1 architecture, several attempts have been reported in the literature. Ho et al. suggested a *Type-1* implementation [19]. Tsay and Newcomb proposed a CMOS circuit technique that would realize a partial *Type-2* implementation [20]; Wunsch et al. [21] have built optical-based *Type-3* implementations; this Chapter presents a CMOS VLSI *Type-3* circuit, [13].

5 A VLSI-Friendly ART1 Algorithm

Let us start describing the *Type-3* model of the original ART1 architecture. The ART1 topology is shown in Figure 1, and consists of two layers: layer *F1* is the input layer and has *N* nodes (one for each binary "pixel" of the input pattern), and layer *F2* is the category layer. Each node in the *F2* layer represents a "cluster" or "category". In this layer only one node will become active after presentation of an input pattern $\mathbf{I} \equiv (I_1, I_2, \ldots, I_N)$. The *F2* layer category that will become active is that which most closely represents the input pattern **I**. If no preexisting category is satisfactory for

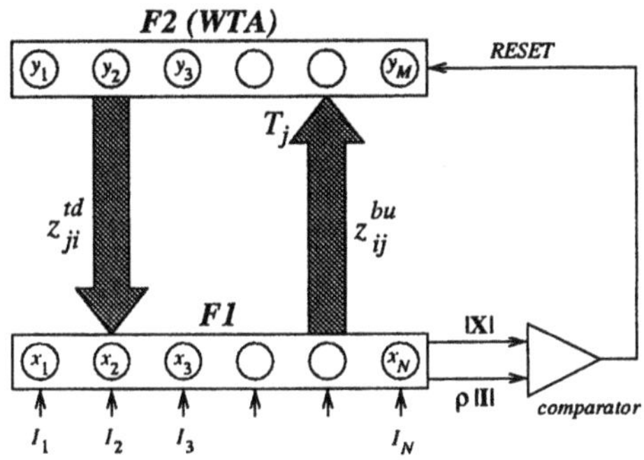

Figure 1. Simplified block diagram of ART1 $Type - 3$ architecture.

a given input pattern, a new category will be formed. Each *F1* node x_i is connected to all *F2* nodes y_j through bottom-up connections of weights z_{ij}^{bu}, so that the input received by each *F2* node y_j is given by

$$T_j = \sum_{i=1}^{N} z_{ij}^{bu} I_i \tag{4}$$

Layer *F2* acts as a Winner-Take-All network, so that all nodes y_j remain inactive, except that which receives the largest bottom-up input T_j,

$$y_j = \begin{cases} 1 & \text{if } T_j = max_k\{T_k\} \\ 0 & \text{otherwise} \end{cases} \tag{5}$$

Once an *F2* winning node arises, a top-down pattern is activated through the top-down weights z_{ji}^{td}. Let us call this top-down pattern $\mathbf{X} \equiv (X_1, X_2, \ldots, X_N)$. The resulting vector $\mathbf{X}$ is given by the equation,

$$X_i = I_i \sum_j z_{ji}^{td} y_j \tag{6}$$

Since only one y_j is active, let us call this winning *F2* node y_J, so that $y_j = 0$ if $j \neq J$ and $y_J = 1$. In this case we can state

$$X_i = I_i z_{Ji}^{td} \quad \text{or} \quad \mathbf{X} = \mathbf{I} \cap \mathbf{z}_J^{td} \tag{7}$$

where $\mathbf{z}_J^{td} \equiv (z_{1J}^{td}, z_{2J}^{td}, \ldots, z_{NJ}^{td})$. This top-down template will be compared with the original input pattern $\mathbf{I}$ according to a predetermined vigilance criterion, tuned by the vigilance parameter $0 < \rho \leq 1$, so that two alternatives may occur:

1. If $\rho|\mathbf{I}| \leq |\mathbf{I} \cap \mathbf{z}_J^{td}|$ the active category J is accepted and the system weights will be updated to incorporate this new knowledge.

2. If $\rho|\mathbf{I}| > |\mathbf{I} \cap \mathbf{z}_J^{td}|$ the active category J is not valid for the actual value of the vigilance parameter ρ. In this case y_J will be deactivated (reset) making $T_J = 0$, so that another y_j node will become active through the Winner-Takes-All action of the $F2$ layer.

Learning takes place when an active $F2$ node is accepted by the vigilance criterion. The weights will be updated according to the following algebraic equations,

$$z_{iJ}^{bu}|_{new} = \frac{LX_i}{L - 1 + |\mathbf{I} \cap \mathbf{z}_J^{td}|_{old}|} = \frac{LI_i z_{Ji}^{td}|_{old}}{L - 1 + |\mathbf{I} \cap \mathbf{z}_J^{td}|_{old}|} \tag{8}$$
$$z_{iJ}^{td}|_{new} = X_i = I_i z_{Ji}^{td}|_{old}$$

or using vector notation

$$\mathbf{z}_J^{bu}|_{new} = \frac{L\mathbf{I} \cap \mathbf{z}_J^{td}|_{old}}{L - 1 + |\mathbf{I} \cap \mathbf{z}_J^{td}|_{old}|} \tag{9}$$
$$\mathbf{z}_J^{td}|_{new} = L\mathbf{I} \cap \mathbf{z}_J^{td}|_{old}$$

where parameter L has to be larger than '1' [12], [13]. Note that only the weights of the connections touching the $F2$ winning node y_J are updated. Therefore, operation of the *Type-3* (or Fast Learning) implementation of the ART1 architecture is described by the algorithm depicted in Figure 2(a).

From a hardware implementation point of view, one of the first issues that comes into consideration is that there are two templates of weights to be built. The set of bottom-up weights z_{ij}^{bu}, each of which must store a real value belonging to the interval[2] [0,K], and the set of top-down weights z_{ji}^{td}, each of which stores either the value '0' or '1'. The physical implementation of the bottom-up template memory presents the first hardware difficulty, because their weights need either an analog or a digital memory with sufficient bits per weight so that the digital discretization does

[2] $K = \frac{L}{L-1+N}$ with $L > 1$ [12], [13]

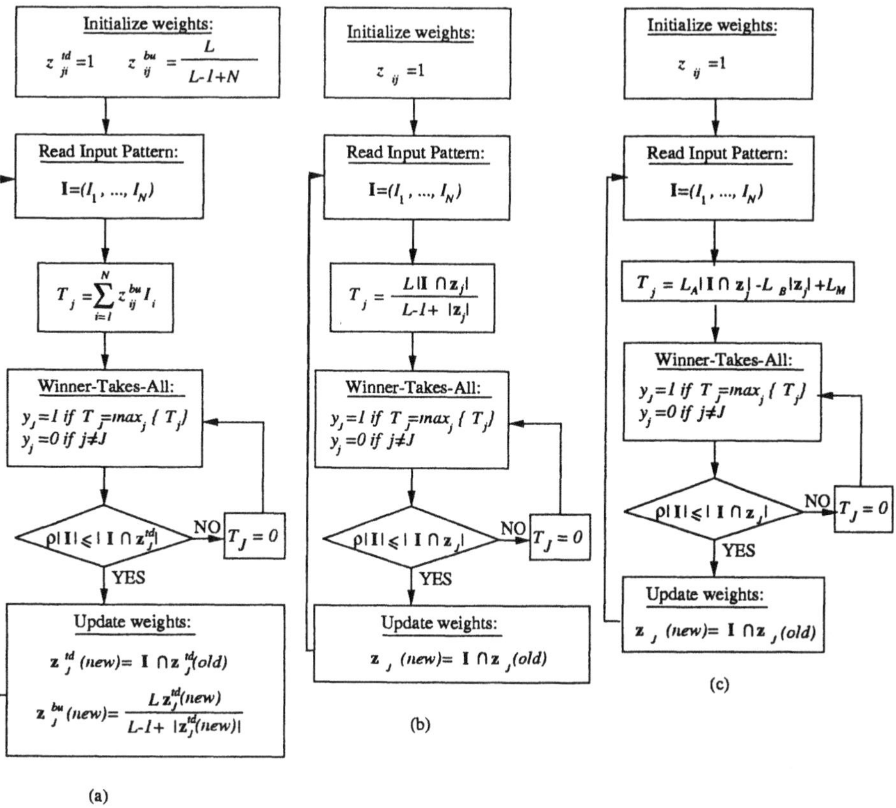

Figure 2. $Type3$ implementation algorithms of the ART1 architecture: (a) original ART1, (b) ART1 with a single valued weights template, (c) modified VLSI-friendly ART1.

not affect the system performance. However, looking at eqs. (9) it can be seen that the bottom-up set $\{z_{ij}^{bu}\}$ and the top- down set $\{z_{ji}^{td}\}$ contain the same information: each of these sets can be fully computed by knowing the other set. It can be seen that the bottom-up set z_{ij}^{bu} is a normalized version of the top-down set z_{ji}^{td}. Therefore, from a hardware implementation point of view it would be desirable to physically implement only a binary valued set (one bit per weight) and let the hardware do the normalization of the bottom-up weights during the computation of $\{T_j\}$. This way, the two sets $\{z_{ij}^{bu}\}$ and $\{z_{ji}^{td}\}$ can be substituted by a single binary valued set $\{z_{ij}\}$, and eq. (4) modified to take into account the normalization effect of the original bottom-up weights,

$$T_j = \frac{LT_{Aj}}{L-1+T_{Bj}} = \frac{L|\mathbf{z}_j \cap \mathbf{I}|}{L-1+|\mathbf{z}_j|} \qquad (10)$$

where $T_{Aj} = \sum_{i=1}^{N} z_{ij} I_i$ and $T_{Bj} = \sum_{i=1}^{N} z_{ij}$. Considering this minor "implementation" modification the algorithm of Figure 2(a) would be transformed into that depicted in Figure 2(b). The system level performance of the algorithms described by Figure 2(a) and Figure 2(b) are identical. There is no difference in the behavior between the two diagrams, and the one in Figure 2(b) offers more attractive features from a hardware (as well as software) implementation point of view.

However, in Figure 2(b) an extra division operation, $T_j = (LT_{Aj})/(L - 1 + T_{Bj})$, need be performed for each node in the $F2$ layer. This is an expensive hardware operation and would probably constitute a performance bottleneck in the overall system for both analog and digital circuit implementations. If possible, it would be very desirable to avoid this division operation. In [13], [22] we show that this division operation can be substituted by a substraction operation, while preserving all the computational properties of the original ART1 algorithm. For some sequence of patterns a different behavior can be observed with respect to the original ART1, but the overall clustering behavior is still equivalent. Mathematically, the input to the $F2$ layer is now,

$$T_j = L_A T_{Aj} - L_B T_{Bj} + L_M \tag{11}$$

where L_A and L_B are positive parameters that play the role of the original L (and $L - 1$) parameter. The condition $L_A > L_B$ must be imposed for proper system operation [22], [13]. $L_M > 0$ is a constant parameter needed to assure that $T_j \geq 0$, for all possible values of T_{Aj} and T_{Bj}.

Replacing a division operation with a substraction one is a very important hardware simplification with significant performance improvement potential. Figure 2(c) shows the final VLSI-friendly *Type-3* ART1 algorithm, which has been mapped into hardware, as described in the next section.

6 Circuit Description

Figure 2(c) shows the VLSI-friendly algorithmic ART1 description, discussed in the previous section, which we want to map into hardware. The operations in Figure 2(c) that need to be implemented are the following:

- Generation of the terms T_j or "choice functions". Since z_{ij} and I_i are binary valued (0 or 1), "binary multiplication" and analog addition/subtraction operations are required.
- Winner-Takes-All (WTA) operation to select the maximum T_j term.
- Comparison of the term $\rho|\mathbf{I}|$ with $|\mathbf{I} \cap \mathbf{z}_J|$.
- Deselection of terms T_J for which $\rho|\mathbf{I}| > |\mathbf{I} \cap \mathbf{z}_J|$.
- Update of weights.

The first three operations require a certain amount of precision (they are analog operations), while the last two operations are not precise (they are digital operations). A reasonable precision for a massive parallel analog circuit implementation is around 1 or 2% (equivalent to 6-bits), while handling input patterns with N of the order of hundreds. Figure 3 shows

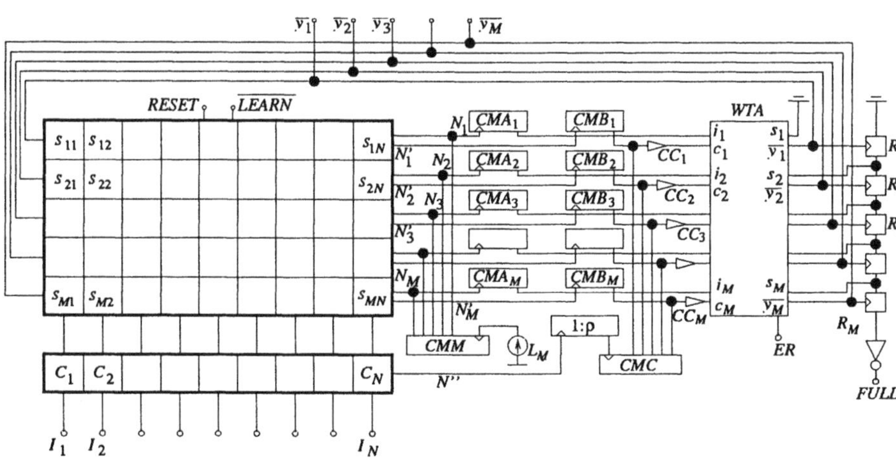

Figure 3. Hardware block diagram for the ART1m algorithm.

a possible hardware block diagram that would physically implement the algorithm of Figure 2(c). The circuit consists of an $M \times N$ array of synapses $S_{11}, S_{12}, \ldots, S_{MN}$, a $1 \times N$ array of controlled current sources $C_1, C_2, \ldots C_N$, two $1 \times M$ arrays of unity-gain current mirrors CMA_1, $\ldots, CMA_M, CMB_1, \ldots, CMB_M$, a $1 \times M$ array of current comparators $CC_1, \ldots, CC_M$, an M-input WTA circuit, two M-output unity-gain current mirrors CMM and CMC, and an adjustable-gain ($0 < \rho \leq 1$)

current mirror. Registers $R_1, \ldots, R_M$ are optional, and their function is explained later.

Each synapse receives two input signals $\overline{y_j}$ and I_i, has two global control signals $RESET$ and $\overline{LEARN}$, stores the value of z_{ij}, and generates two output currents:

- the first goes to the input of current mirror CMA_j (node N_j) and is $L_A z_{ij} I_i - L_B z_{ij}$.
- the second goes to the input of current mirror CMB_j (node $N_j^\prime$) and is $L_A z_{ij} I_i$.

All synapses in the same row j ($S_{j1}, S_{j2}, \ldots, S_{jN}$) share the two nodes ($N_j$ and N'_j) into which the currents they generate are injected. Therefore, the input of current mirror CMA_j receives the current

$$T_j = L_A \sum_{i=1}^{N} z_{ij} I_i - L_B \sum_{i=1}^{N} z_{ij} + L_M = L_A |\mathbf{I} \cap \mathbf{z}_j| - L_B |\mathbf{z}_j| + L_M \quad (12)$$

while the input of current mirror CMB_j receives the current

$$L_A \sum_{i=1}^{N} z_{ij} I_i = L_A |\mathbf{I} \cap \mathbf{z}_j| \quad (13)$$

Current L_M, which is replicated M times by current mirror CMM has an arbitrary value as long as it assures that all terms T_j are positive.

Each element of the array of controlled current sources C_i has one binary input signal I_i and generates the current $L_A I_i$. All elements C_i share their output node N'', so that the total current they generate is $L_A |\mathbf{I}|$. This current reaches the input of the adjustable gain ρ current mirror, and is later replicated M times by current mirror CMC.

Each of the M current comparators CC_j receives the current $L_A |\mathbf{I} \cap \mathbf{z}_j| - L_A \rho |\mathbf{I}|$ and compares it against zero (i.e. it checks the vigilance criterion for category j). If this current is positive, the output of the current comparator falls, but if the current is negative the output rises. Each current comparator CC_j output controls input c_j of the WTA. If c_j is high the current sunk by the WTA input i_j (which is T_j) will not compete for the winning node. On the contrary, if c_j is low, input current T_j will enter the WTA competition. The outputs of the WTA $\overline{y_j}$ are all high, except for that which receives the largest $\overline{c_j}T_j$: such output, denominated $\overline{y_J}$, will fall.

Now we can describe how the operation of the circuit in Figure 3 follows that of the flow diagram of Figure 2(c). All synaptic memory values z_{ij} are initially set to '1' by the $RESET$ signal. Once the input vector $\mathbf{I}$ is activated, the M rows of synapses generate the currents $L_A|\mathbf{I} \cap \mathbf{z}_j| - L_B|\mathbf{z}_j|$ and $L_A|\mathbf{I} \cap \mathbf{z}_j|$, and the row of controlled current sources $C_1, \ldots, C_N$ generates the current $L_A|\mathbf{I}|$. Each current comparator CC_j will prevent current $T_j = L_A|\mathbf{I} \cap \mathbf{z}_j| - L_B|\mathbf{z}_j| + L_M$ from competing in the WTA if $\rho|\mathbf{I}| > |\mathbf{I} \cap \mathbf{z}_j|$. Therefore, the effective WTA inputs are $\{\overline{c_j}T_j\}$, from which the WTA chooses the maximum, making the corresponding output $\overline{y_J}$ fall. Once $\overline{y_J}$ falls, the synaptic control signal $\overline{LEARN}$ is temporarily set low, and all z_{iJ} values will change from '1' to 'I_i'.

Note that initially (when all $z_{ij} = 1$),

$$\overline{c_j}T_j = L_A|\mathbf{I}| - L_B N + L_M \quad \forall j \tag{14}$$

This means that the winner will be chosen among M equal competing inputs, basing the election on mismatches due to random process parameter variations of the transistors. Even after some categories are learned, there will be a number of uncommitted rows ($z_{1j} = \cdots = z_{Nj} = 1$) that generate the same competing current of eq. (14). The operation of a WTA circuit in which there are more than 1 equal and winning inputs becomes more difficult and in the best case, renders slower operation. To avoid these problems M D-registers, $R_1, \ldots, R_M$, might be added. Initially these registers are set to '1' so that the WTA inputs $s_2, \ldots, s_M$ are high. Inputs $s_1, \ldots, s_M$ have the same effect as inputs $c_1, \ldots, c_M$: if s_j is high T_j does not compete for the winner, but if s_j is low, T_j enters the WTA competition. Therefore, initially only $\overline{c_1}T_1$ competes for the winner. As soon as $\overline{y_1}$ falls once, the input of register R_1 (which is '0') is transmitted to its output making $s_2 = 0$. Now both $\overline{c_1}T_1$ and $\overline{c_2}T_2$ will compete for the winner. As soon as $\overline{c_2}T_2$ wins once, the input of register R_2 is transmitted to its output making $s_3 = 0$. Now $\overline{c_1}T_1$, $\overline{c_2}T_2$, and $\overline{c_3}T_3$

will compete, and so on. If all available $F2$ nodes $(y_1, \ldots, y_M)$ have won once, the $FULL$ signal rises, advising that all $F2$ nodes are storing a category. The WTA control signal ER enables operation of these registers. Next we describe the circuit implementation for each block in Figure 3.

6.1 Synaptic Circuit and Controlled Current Sources

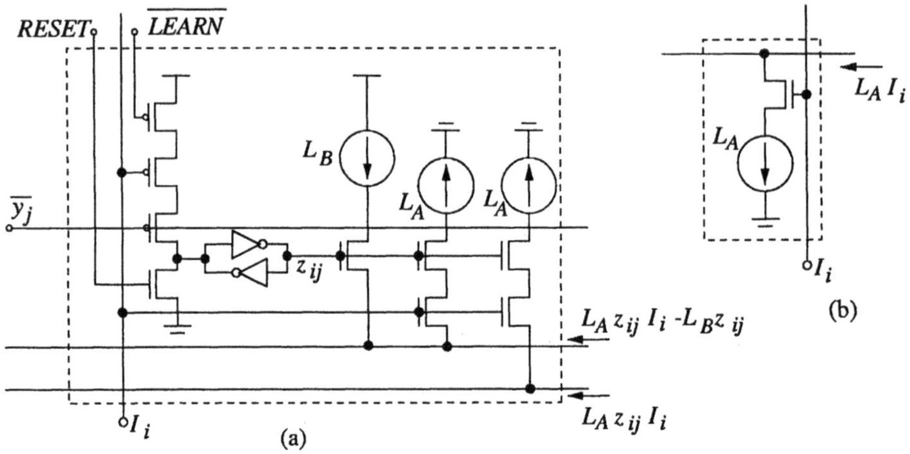

Figure 4. (a) Details of synaptic circuit S_{ij}, (b) Details of controlled current source circuit C_i.

The details of a synapse S_{ji} are shown in Figure 4(a). It consists of three current sources (two of value L_A and one of value L_B), a two-inverter loop (acting as a Flip-Flop), and nine MOS transistors working as switches. As can be seen in Figure 4(a) each synapse generates the currents $L_A z_{ij} I_i - L_B z_{ij}$ and $L_A z_{ij} I_i$. The $RESET$ control signal sets z_{ij} to '1'. Learning is performed by making z_{ij} change from '1' to '0' whenever $\overline{LEARN} = 0$, $\overline{y_j} = 0$, and $I_i = 0$.

Figure 4(b) shows the details of each controlled current switch C_i. If $I_i = 0$ no current is generated, while if $I_i = 1$, the current L_A is provided.

6.2 Winner-Takes-All (WTA) Circuit

Figure 5 shows the details of the WTA circuit. It is based on Lazzaro's WTA [24], which consists of the array of transistors MA and MB, and

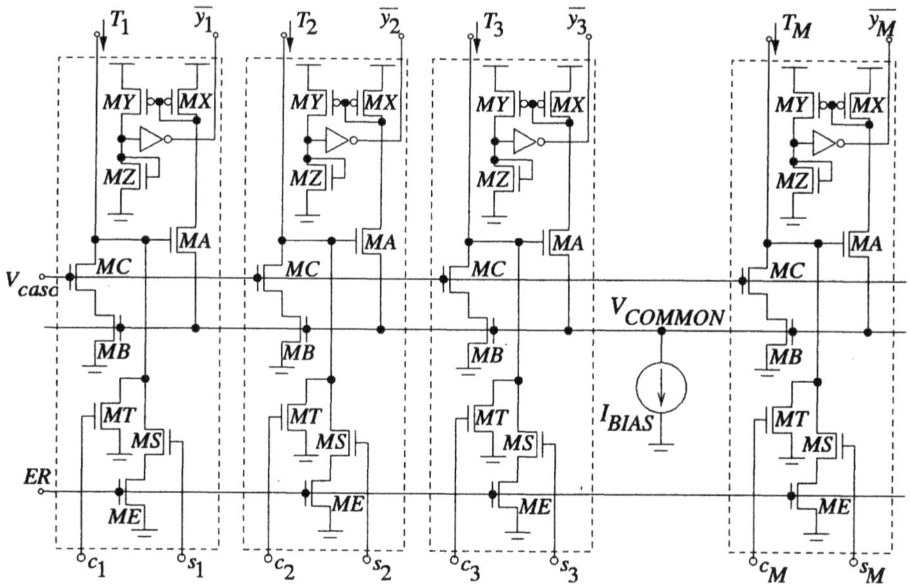

Figure 5. Circuit schematic of Winner-Takes-All (WTA) circuit.

the current source I_{BIAS}. Transistor MC has been added to introduce a cascade effect and increase the gain of each cell. Transistors MX, MY, and MZ transform the output current into a voltage, which is then inverted to generate $\overline{y_j}$. Transistor MT disables the cell if c_j is high, so that the input current T_j will not compete for the winner. Transistors MS and ME have the same effect as transistor MT: if signals ER and s_j are high, T_j will not compete.

6.3 Current Comparators

The circuit used for the current comparators is shown in Figure 6(a). Such a comparator forces an input voltage approximately equal to the inverters trip voltage, has extremely high resolution (less than $1pA$), and can be extremely fast (in the order of $10 - 20ns$ for input around $10\mu A$) [25]. Note that in the steady state the inverter is biased in its trip-point, and therefore is consuming a significant amount of current. If power consumption is of concern this is not acceptable. Instead, the comparator of Figure 6(b) should be used which uses an OTA [26] as voltage amplifier.

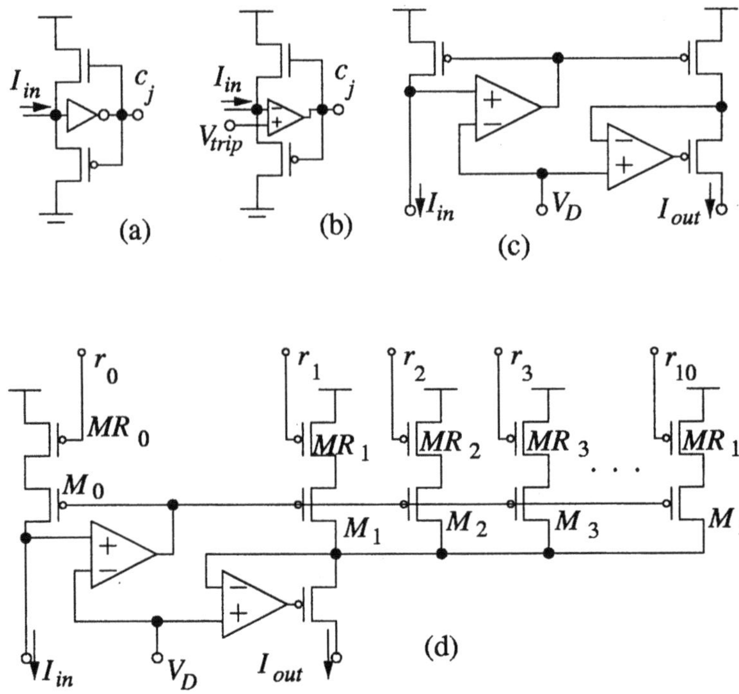

Figure 6. (a) Circuit schematic of current comparator using digital inverter or (b) OTA, (c) of active-input regulated-cascadecurrent mirror, (d) and of adjustable gain ρ current mirror.

6.4 Current Mirrors

Current mirrors CMA_j, CMB_j, CMM, CMC, and the ρ-gain mirror have been laid out using common centroid layout techniques to minimize matching errors and keep a 6-bit precision for the overall system. For current mirrors CMA_j and CMB_j a special topology has been used, shown in Figure 6(c) [27], [13]. This topology forces a constant voltage V_D at its input and output nodes, thus producing a virtual ground in the output nodes of all synapses, which reduces channel length modulation distortion improving matching between the currents generated by all synapses. In addition, the topology of Figure 6(c) presents a very wide current range with small systematic (non-random) matching errors [13], [27].

The adjustable gain ρ current mirror also uses this topology, as shown in Figure 6(d). Transistor M_0 has a geometry factor (W/L) 10 times larger than transistors $M_1, \ldots, M_{10}$. Transistors $MR_1, \ldots, MR_{10}$ act as switches (controlled by signals $r_1, \ldots, r_{10}$), so that the gain of the current mirror can be adjusted between $\rho = 0.0$ to $\rho = 1.0$ in steps of 0.1, while maintaining $r_0 = 0$. By making r_0 slightly higher than 0 Volts, ρ can be fine tuned.

7 Modular System Expansivity

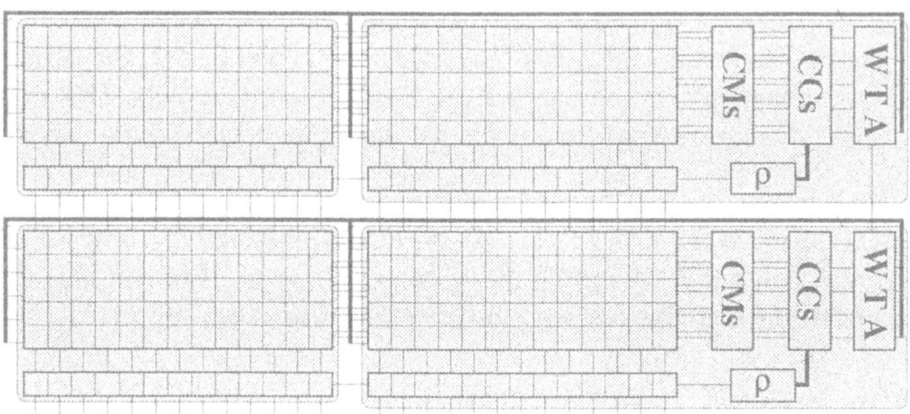

Figure 7. Interchip connectivity for modular system expansion.

The circuit of Figure 3 can be expanded both horizontally, increasing the number of input patterns from N to $N \times n_H$, and vertically increasing the number of possible categories from M to $M \times n_V$, just by assembling an array of $n_H \times n_V$ chips. Figure 7 shows schematically the interconnectivity between chips in the case of a 2×2 array.

Vertical expansion of the system is possible by making several chips share the input vector terminals $I_1, \ldots, I_N$, and node V_{COMMON} of the WTA (see Figure 5). Thus, the only requirement is that V_{COMMON} be externally accessible. Horizontal expansion is directly possible by making all chips in the same row share their N_j, N'_j, and N'' nodes, and isolating all except one of them, from the current mirrors CMA_j, CMB_j, and the adjustable gain ρ-mirror. Also, all synapse inputs $\overline{y_j}$ must be shared.

Both vertical and horizontal expansion degrades the system performance. Vertical expansion causes degradation because the WTA becomes distributed among several chips. For the WTA of Figure 5, all MA transistors must match well, which is very unlikely if they are in different chips. A solution for this problem is to use a WTA circuit technique based on current processing and replication, insensitive to inter-chip transistor mismatches [13], [28], [29].

Horizontal expansion degrades the performance because current levels have to be changed:

- Either currents L_A and L_B are maintained the same, which makes the current mirrors CMA_j, CMB_j, CMM, $1:\rho$, and CMC, the current comparators CCj, and the WTA to handle higher currents. This may cause malfunctioning due to eventual saturation in some of the blocks.
- Or currents L_A and L_B are scaled down so that the current mirrors CMA_j, CMB_j, CMM, $1:\rho$, and CMC, the current comparators CCj, and the WTA handle the same current level. However, this degrades the matching among synaptic current sources [30], [31].

Also, horizontal expansion requires to access externally $3M$ lines for each chip (nodes $\overline{y_j}$, N_j, and N_j'). If M is large (of the order of hundreds) horizontal expansion becomes unfeasible.

8 Implementation of Synaptic Current Sources

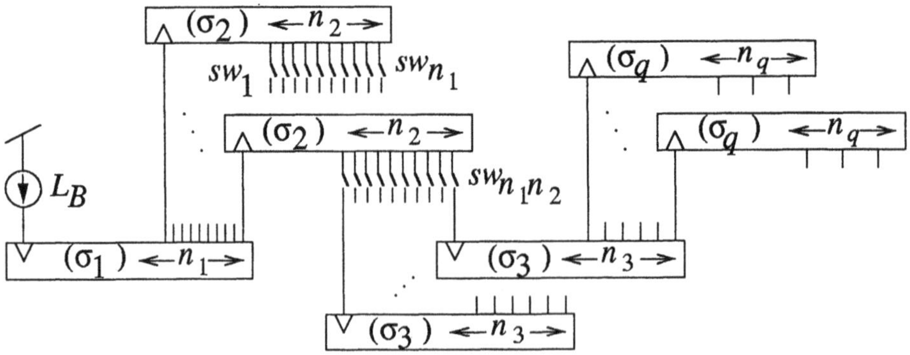

Figure 8. Cascade of current mirrors for low mismatch.

The physical implementation of the *Synaptic Current Sources* is the one that most critically affects the final density of the resulting system as well as its precision.

The most simple way would be to build a current mirror with multiple outputs: for example, for current sources L_B, one can make a current mirror with $N \times M$ output transistors and have each of them placed in a synapse. For the L_A current sources a mirror with $2N \times M$ outputs would be needed. The problem with this implementation is that the distance between input and output transistors can become extremely large and therefore produce a very large mismatch between input and output currents. For example, if we want to make an array of size around $1cm^2$ using $40\mu m \times 40\mu m$ transistors, biased with currents of $10\mu A$, a worst case mismatch of 10% would be optimistic. However, if instead of distances of the order of $1cm$, we can spread an array of transistors over distances of the order of $1mm$, we might be able to achieve mismatches below 1%. The Appendix explains a cheap and easy methodology for characterizing transistor mismatch of a given technology. This methodology also allows us to identify approximately what is the maximum die area over which we can spread an array of transistors (like the outputs of a current mirror) while maintaining mismatch below or close to 1%.

As opposed to large transistor arrays, a single current mirror, with a reduced number of outputs (like 10), a reasonable transistor size (like $40\mu m \times 40\mu m$), a moderate current (around $10\mu A$), and using common centroid layout techniques can be expected to have a mismatch error standard deviation σ_q of less than 1% [30], [31]. By cascading several of these current mirrors in a tree-like fashion as is shown in Figure 8 (for current sources L_B), a high number of current sources (copied from a single common reference) can be generated with a mismatch characterized by

$$\sigma_{Total}^2 = \sigma_1^2 + \sigma_2^2 + \cdots + \sigma_q^2 \tag{15}$$

Each current mirror stage introduces a random error of standard deviation σ_k. This error can be reduced by increasing the transistor areas of the current mirrors. Since the last stage q has a higher number of current mirrors, it is important to keep their area low. For previous stages the transistors can be made larger to contribute with a smaller σ_k, because they are less in number and will not contribute significantly to the total transistor area. For current sources L_A, a circuit similar to that shown in

Figure 8 can be used. Current L_B in Figure 8 (and similarly current L_A) is injected externally into the chip so that parameter $\alpha = L_A/L_B$ can be controlled.

In the next sections we provide experimental results related to two ART1 chip implementations. They differ only in the way the L_A and L_B current sources are implemented. In the first prototype they were implemented using the current mirror tree approach[3] of Figure 8. This produces a very high area consumption (since each stage, including the last, is laid out using common centroid structures). We were able to fit an array of 100×18 synapses into a $1cm^2$ die, for a $1.6\mu m$ CMOS technology.

The second prototype is made by directly making one current mirror with $N \times M$ outputs (for L_B) and another with $2N \times M$ outputs (for L_A), but limiting the array to $2.75mm^2$. We were able to fit an array of 50×10 synapses into this area, for a $1.0\mu m$ CMOS technology.

The optimum strategy for building large arrays would be to tile a large area (like $1cm^2$) into pieces of smaller areas (like $1 - 2mm^2$). In each tile one can implement the current mirrors with a very large number of outputs (as in the approach of the second prototype), and feed the input of each tile current mirror using a tree-mirror structure (as in the first prototype). This approach is schematically depicted in Figure 9. When drawing the layout, care should be taken to assure that each synapse has a similar neighborhood whether or not it lies on the edge of a tile or not. This would avoid matching degradation due to layout neighborhood effects [32].

8.1 Weights Read Out

A very important feature for test purposes is to be able to read out the weight z_{ij} for each synapse in the array. In the first prototype, the switches sw_1 to $sw_{n_1 \times n_2}$ of Figure 8 were added to enable z_{ij} read out column by column and test the progress of the learning algorithm. These switches are all ON during normal operation of the system. However, for weights read-out, all except one will be OFF. The switch that is ON

[3]The chip was designed in this way because at that time no information on transistor mismatch over large distances ($1cm$) was available.

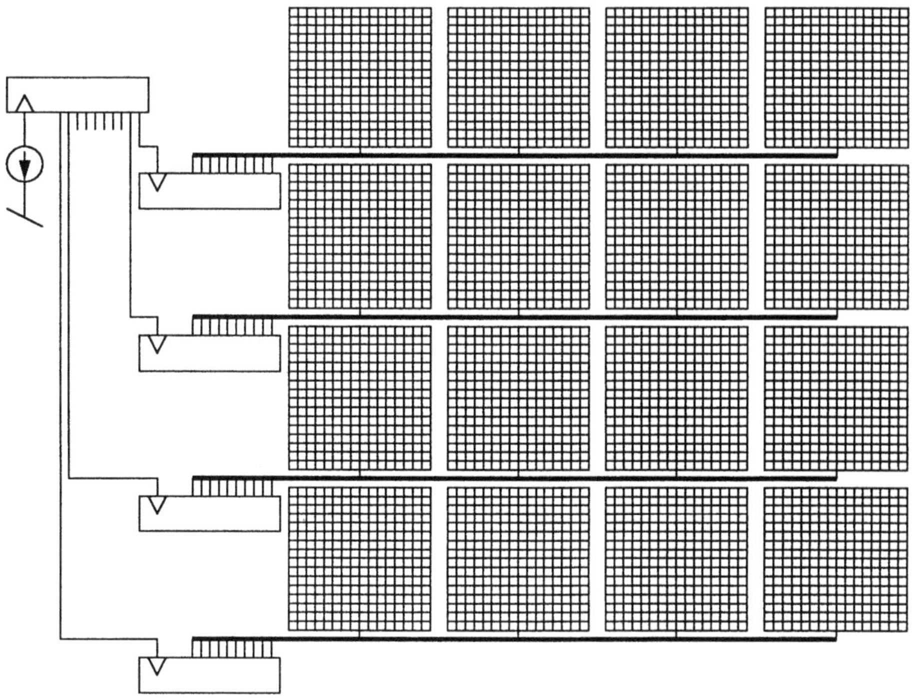

Figure 9. Optimum strategy for minimum current sources mismatch for large area arrays.

is selected by a decoder inside the chip, so that only column i of the synaptic array of Figure 3 injects the current $z_{ij}L_B$ to nodes N_j. During weights read out, either current L_A is set to zero or input pattern $\mathbf{I} \equiv \mathbf{0}$ is loaded, so that only the L_B current sources inject current into nodes N_j. All nodes N_j can be isolated from current mirrors CMA_j, and connected to output pads to sense the currents $z_{ij}L_B$.

For the second prototype an extra MOS switch was added in series with each current source L_A and L_B. This switch is controlled by a common line for each column of the synaptic array. During normal operation all these switches are ON, but during weights read out all are OFF except for one column. The column is selected through an on-chip decoder.

9 Experimental Results of First Prototype

Following the approach of Figure 8, a prototype chip was fabricated in a standard single-poly double-metal $1.6\mu m$ CMOS digital process (Eurochip ES2). The die area is $1cm^2$ and it has been mounted in a 120-pin PGA package. This chip implements an ART1 system with 100 nodes in the $F1$ layer and 18 nodes in the $F2$ layer. Most of the pins are intended for test and characterization purposes. All the subcircuits in the chip can be isolated from the rest and conveniently characterized. The $F1$ input vector **I**, which has 100 components, has to be loaded serially through one of the pins into a shift register. The time delay measurements reported next do not include the time for loading this shift register.

The experimental measurements provided in this section have been divided into four parts. The first describes DC characterization results of the elements that contribute critically to the overall system precision. These elements are the WTA circuit and the synaptic current sources. The second describes time delay measurements that contribute to the global throughput time of the system. The third presents system level experimental behaviors obtained with digital test equipment (HP82000). Finally, the fourth focuses on yield and fault tolerance characterizations.

9.1 System Precision Characterizations

The ART1 chip was intended to achieve an equivalent 6-bit (around 1.5% error) precision. The part of the system that is responsible for the overall precision is formed by the components that perform analog computations. These components are (see Figure 3) all current sources L_A and L_B, all current mirrors CMA_j, CMB_j, CMM, $1 : \rho$, CMC, the current comparators CC_j, and the WTA circuit. The most critical of these components (in precision) is the WTA circuit. Current sources and current mirrors can be made to have mismatch errors below 0.2% [30], [31], [33], [34], [35], [Appendix] at the expense of increasing transistors area and current, decreasing distances between matched devices, and using common centroid layout techniques [36]. This is feasible for current mirrors CMA_j, CMB_j, CMM, $1 : \rho$, and CMC, which appear in small numbers. However, the area and current level is limited for the synaptic current sources L_A and L_B, since there are many of them. Therefore,

Table 1. Precision of the WTA.

T_j	$10\mu A$	$100\mu A$	$1mA$
$\sigma(T_j)$	1.73%	0.86%	0.99%

WTA and current sources L_A and L_B are the elements that limit the precision of the overall system, and their characterization results will be described next.

9.1.1 WTA Precision Measurements

L_A and L_B will have current values of $10\mu A$ or less. The maximum current a WTA input branch can receive is (see eq. (12)),

$$T_j|_{max} = L_M + \left[\sum_{i=1}^{100} z_{ij}(L_A I_i - L_B) \right]_{max} = L_M + 100(L_A - L_B) \quad (16)$$

which corresponds to the case where all z_{ij} and I_i values are equal to '1' (remember that $L_A > L_B > 0$). In our circuit the WTA was designed to handle input currents of up to $1.5mA$ for each input branch. In order to measure the precision of the WTA, all input currents except two were set to zero. Of these two inputs one was set to $100\mu A$ and the other was swept between $98\mu A$ and $102\mu A$. This will cause their corresponding output voltages $\overline{y_j}$ to indicate an interchange of winners. The transitions do not occur exactly at $100\mu A$. Moreover, the transitions change with the input branches. The standard deviation of these transitions was measured as $\sigma = 0.86\mu A$ (or 0.86%). Table 1 shows the standard deviation (in %) measured when the constant current is set to $10\mu A$, $100\mu A$, and $1mA$.

9.1.2 Synaptic Current Sources Precision Measurements

The second critical precision error source of the system is the mismatch between synaptic current sources. In the chip each of the $N \times (2M + 1)$ L_A current sources and each of the $N \times M$ L_B current sources could be isolated and independently characterized. Figure 10 shows the measured mismatch error (in %) for 18 arbitrary L_A current sources when sweeping their currents between $0.1\mu A$ and $10\mu A$. As can be seen in Figure 10, for currents higher than $5\mu A$ the standard deviation of the mismatch error is below 1%. The same result is obtained for the L_B current sources.

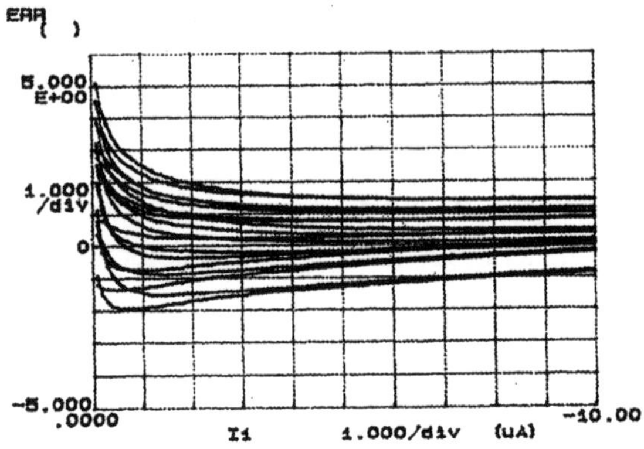

Figure 10. Measured mismatch error (in %) between 18 arbitrary L_A current sources.

9.2 Throughput Time Measurements

For a real-time clustering device the throughput time can be defined as the time needed for each input pattern to be processed. During this time the input pattern has to be classified into one of the pre-existing categories or assigned to a new one, and the pre-existing knowledge of the system has to be updated to incorporate the new information the input pattern carries. From a circuit point of view, this translates into the measurement of two delay times:

1. The time needed by the WTA to select the maximum among all $\overline{c_j}T_j$.
2. The time needed by the synaptic cells to change z_{ij} from its old value to $y_j I_i z_{ij}$.

9.2.1 WTA Delay Measurements

The delay introduced by the WTA depends on the current level present in the competing input branches. This current level will depend on the values chosen for L_A, L_B, and L_M, as well as on the input pattern $\mathbf{I}$ and all internal weights $\mathbf{z}_j$. To keep the presentation simple, delay times will be given as a function of T_j values directly. Table 2 shows the measured delay times when T_1 changes from T_1^a to T_1^b, and T_2 to T_{18} have the values given in the table. Delay t_{d1} is the time needed by category y_1 to win when T_1 switches from T_1^a to T_1^b, and t_{d2} is the time needed by

Table 2. Delay times of the WTA.

T_1^a	T_1^b	T_2	$T_3, \ldots T_{18}$	t_{d1}	t_{d2}
$0\mu A$	$200\mu A$	$100\mu A$	0	$550ns$	$570ns$
$0\mu A$	$1mA$	$500\mu A$	0	$210ns$	$460ns$
$100\mu A$	$150\mu A$	$125\mu A$	$100\mu A$	$660ns$	$470ns$
$400\mu A$	$600\mu A$	$500\mu A$	$400\mu A$	$440ns$	$400ns$
$500\mu A$	$1.50mA$	$1.00mA$	$500\mu A$	$230ns$	$320ns$
$90\mu A$	$110\mu A$	$100\mu A$	0	$1.12\mu s$	$1.11\mu s$
$490\mu A$	$510\mu A$	$500\mu A$	0	$1.19\mu s$	$1.06\mu s$
$990\mu A$	$1.01mA$	$1.00mA$	0	$380ns$	$920ns$

category y_2 to win when T_1 decreases from T_1^b to T_1^a. As can be seen, this delay is always below $1.2\mu s$.

For the cases when the vigilance criterion is not directly satisfied and hence comparators CC_j cut some of the T_j currents, an additional delay is observed. This extra delay has been measured to be less than $400ns$ for the worst cases. Therefore, the time needed until the WTA selects the maximum among all $\overline{c_j}T_j$ is less than $1.2\mu s + 0.4\mu s = 1.6\mu s$.

9.2.2 Learning Time

After a delay of $1.6\mu s$ (so that the WTA can settle), the learn signal $\overline{LEARN}$ (see Figure 3) is enabled during a time t_{LEARN}. To measure the minimum t_{LEARN} time required, this time was set to a specific value during a training/learning trial, and it was checked that the weights had been updated properly. By progressively decreasing t_{LEARN} until some of the weights did not update correctly, it was found that the minimum time for proper operation was $190ns$. By setting t_{LEARN} to $200ns$ and allowing the WTA a delay of $1.6\mu s$, the total throughput time of the ART1 chip is established as $1.8\mu s$.

9.3 Comparison with Digital Neural Processors

A digital chip with a feedforward speed of a connections per second, a learning speed of b connection updates per second, and a WTA section with a delay of c seconds must satisfy the following equation to achieve a throughput time of $1.8\mu s$ when emulating the ART1 algorithm of Figure 3(c):

$$\frac{N \times (2M+1)}{a} + \frac{N}{b} + c = 1.8\mu s \tag{17}$$

Note that there are N synapse weights z_{ij} to update for each pattern presentation, and $N(2M+1)$ feed-forward connections: $N \times M$ connections to generate all terms $T_j = L_A|\mathbf{I} \cap \mathbf{z}_j| - L_B|\mathbf{z}_j| + L_M$, $N \times M$ connections to generate terms $L_A|\mathbf{I} \cap \mathbf{z}_j|$, and N connections to generate $L_A|\mathbf{I}|$.

Assuming $c = 100ns$, and $a = b$, eq. (17) results in a processing speed of $a = b = 2.2 \times 10^9$ connections/s or connection-updates/s for $N = 100$ and $M = 18$. A digital neural processor would require such figures of merit to equal the processing time of the analog ART1 chip presented here. Therefore, this "approximate reasoning" makes us conclude that this chip has an equivalent computing power of $a + b = 4.4 \times 10^9$ connections/s plus connection-updates/s.

9.4 System Level Performance

Although the internal processing of the chip is analog in nature, its inputs (I_i) and outputs ($\overline{y_j}$) are binary valued. Therefore, the system level behavior of the chip can be tested using conventional digital test equipment. In our case we used the HP82000 IC Evaluation System.

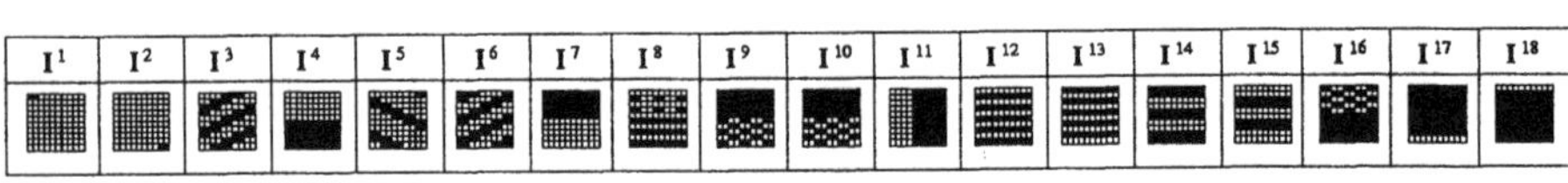

Figure 11. Set of input patterns.

An arbitrary set of 100-bit input patterns $\mathbf{I}^k$ was chosen, shown in Figure 11. A typical clustering sequence is shown in Figure 12, for $\rho = 0.7$ and $\alpha = L_A/L_B = 1.05$. The first column indicates the input pattern $\mathbf{I}^k$ that is fed to the $F1$ layer. The other 18 squares (10×10 pixels) in each row represent each of the internal $\mathbf{z}_j$ vectors after learning is finished. The vertical bars to the right of some $\mathbf{z}_j$ squares indicate that these categories won the WTA competition while satisfying the vigilance criterion. Therefore, such categories correspond to $\mathbf{z}_J$, and these are the only ones that are updated for that input pattern $\mathbf{I}^k$ presentation. The figure shows only two iterations of input patterns presentation, because no

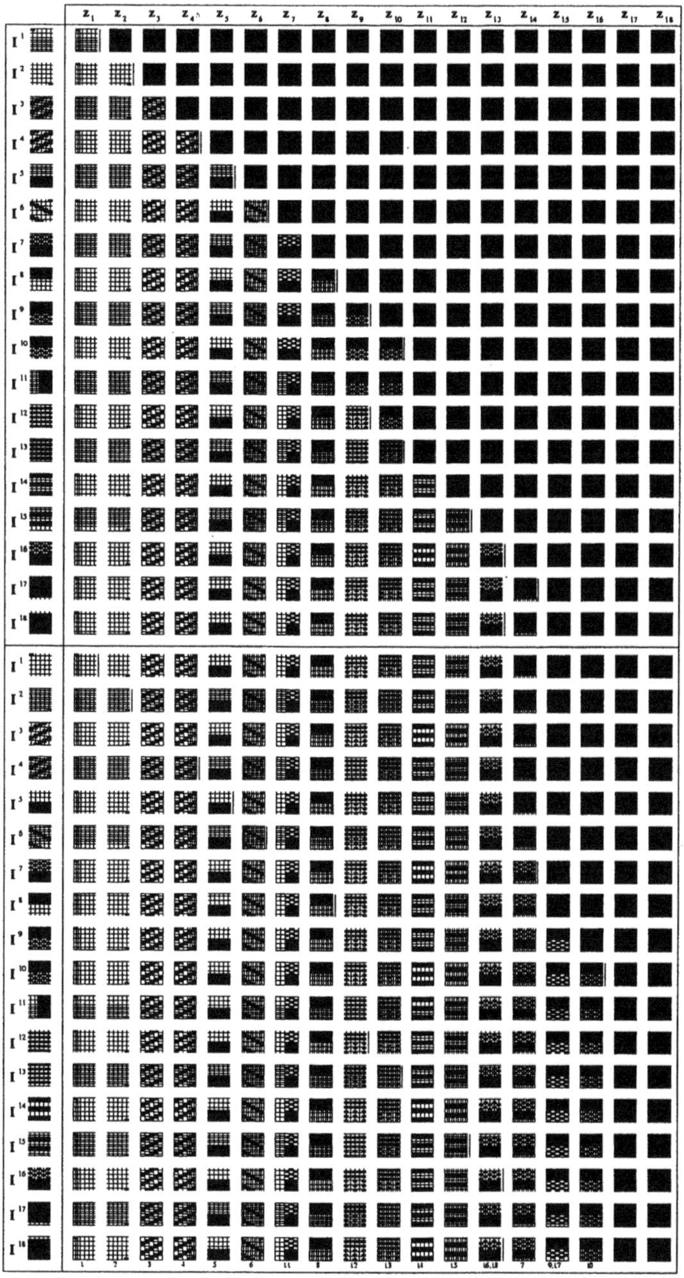

Figure 12. Clustering sequence for $\rho = 0.7$ and $\alpha = L_A/L_B = 1.05$.

change in weights were observed after these. The last row of weights $\mathbf{z}_j$ indicates the resulting categorization of the input patterns. The numbers below each category indicate the input patterns that have been clustered into this category. In the following figures we will show only this last row of learned patterns together with the pattern numbers that have been clustered into each category.

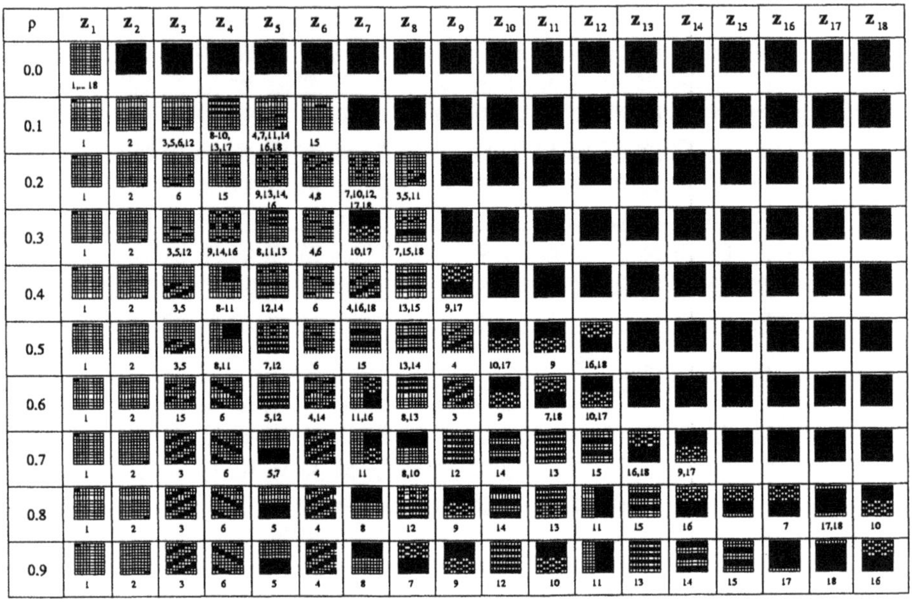

Figure 13. Categorization of the input patterns for $\alpha = 1.07$ and different values for ρ.

Figure 13 shows the categorizations that result when tuning the vigilance parameter ρ to different values while the currents were set to $L_A = 3.2\mu A$, $L_B = 3.0\mu A$, and $L_M = 400\mu A$ ($\alpha = L_A/L_B = 1.07$). Note that below some categories there is no number. This is a known ART1 behavior: during the clustering process some categories might be created which will later be abandoned and will not represent any of the training patterns. In Figure 14 the vigilance parameter is maintained constant at $\rho = 0$, while α changes from 1.07 to 50.

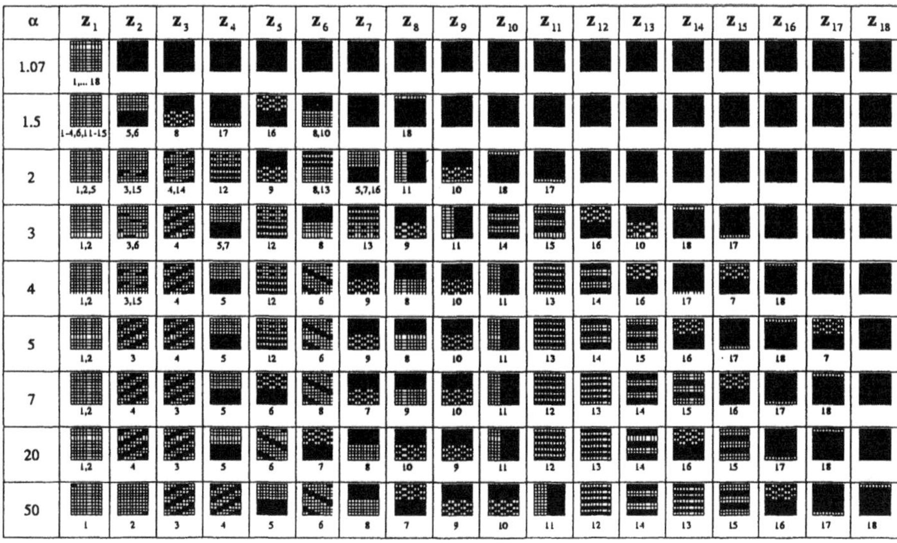

Figure 14. Categorization of the input patterns for $\rho = 0$ and different values of α.

9.5 Yield and Fault Tolerance

A total of 30 chips (numbered 1 through 30 in Table 3 and Figure 15) were fabricated. For each chip every subcircuit was independently tested and its proper operation verified; 14 different faults were identified. Table 3 indicates the faults detected for each of the 30 chips. The faults have been denoted from **F1** to **F14**, and are separated into two groups:

- *Catastrophic Faults* are those clearly originated by a short or open circuit failure. These faults are **F1**, ... **F8**. This kind of faults would produce a failure in a digital circuit.
- *Non-Catastrophic Faults* are those that produce a large deviation from the nominal behavior, too large to be explained by random process parameter variations. These faults are **F9**, ... **F14**. This kind of faults would probably not produce a catastrophic failure in a digital circuit, but be responsible for significant delay times degradations.

Table 4 describes the subcircuits where the faults of Table 3 were found. Note that the most frequent faults are **F3/F9** and **F4/F10**, which are fail-

Table 3. Classification of faults detected in the 30 fabricated chips.

	Catastrophic Faults								Non-Catastrophic Faults					
chip #	F1	F2	F3	F4	F5	F6	F7	F8	F9	F10	F11	F12	F13	F14
1					X						X			
2		X					X		X	X		X	X	
3									X	X		X		X
4		X							X	X		X		
5	X	X							X					
6			X	X					X	X	X			
7			X						X					
8									X					
9		X		X					X	X				
10			X	X					X	X				
11	X													
12				X						X				
13	X	X	X						X	X				
14		X							X					
15		X	X						X	X				
16			X						X					
17	X		X						X					
18									X	X		X		
19										X				
20		X	X	X					X	X		X		
21	X								X					
22									X			X		
23	X								X					
24				X						X				
25														
26									X					
27	X	X	X						X					
28														
29	X													
30									X	X		X		

ures in some current sources L_A or L_B, and these current sources occupy a significant percentage of the total die area. Fault **F1** is a fault in the shift register that loads the input vector $\mathbf{I}^k$. Fault **F2** is a fault in the WTA circuit. Therefore, chips with an **F1** or **F2** fault could not be

Table 4. Description of faults.

F1	non-operative shift register for loading $\mathbf{I}^k$
F2	non-operative WTA circuit
F3/F9	fault in a current source L_A
F4/F10	fault in a current source L_B
F5/F11	fault in $1:\rho$ current mirror
F6/F12	fault in current mirror CMM
F7/F13	fault in current mirrors CMA_j or CMB_j
F8/F14	fault in current mirror CMC

tested for system level operation[4]. Faults **F3** and **F9** are faults detected in the same subcircuits of the chip, with **F3** being catastrophic and **F9** non-catastrophic. The same is valid for **F4** and **F10**, **F5** and **F11**, and so on until **F8** and **F14**.

Note that only 2 of the 30 chips (6.7%) are completely fault-free (chips 25 and 28). According to the simplified expression for the yield performance as a function of die area Ω and process defects density ρ_D [37],

$$\text{yield}(\%) = 100e^{-\rho_D \Omega} \tag{18}$$

this requires a process defect density [5] of $\rho_D = 3.2cm^{-1}$. On the other hand, ignoring the non-catastrophic faults yields 9 out of 30 chips (30%) (chips without catastrophic faults are 3, 8, 18, 19, 22, 25, 26, 28, and 30). According to eq. (18) such a yield would be predicted if the process defect density is $\rho'_D = 1.4cm^{-1}$.

Even though the yield is quite low, many of the faulty samples were still operative. This is due to the fault tolerant nature of the neural algorithms in general [38], [39],[40], and the ART1 algorithm in particular. There were 16 chips with operative shift register and WTA circuit. We performed system level operation tests on these chips to verify if they would be able to form clusters of the input data, and verified that 12 of these 16 chips were able to do so. Moreover, 6 (among which were the two completely fault-free chips) behaved exactly identically. The resulting clustering behavior of these 12 chips is depicted in Figure 15 for $\rho = 0.5$

[4]Chips 9 and 14 had the WTA partially operative and could be tested for system level behavior

[5]The effective die area is $\Omega = (0.92cm)^2$

and $\alpha = 1.07$.

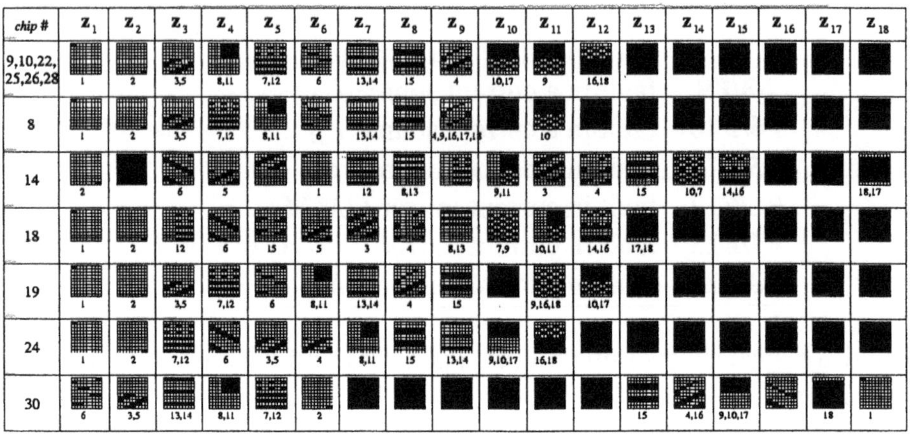

Figure 15. Categorization of the input patterns performed by operative samples.

10 Experimental Results of Second Prototype

A second prototype chip was fabricated in a $1.0\mu m$ CMOS process (ES2 through Eurochip) by implementing the L_A and L_B current sources directly as single multiple-outputs current mirrors [41], [13]. This could be done because at that time we had some information on long distance gradient induced mismatch for this technology. In the Appendix we show how we performed transistor random mismatch characterization using a special purpose chip for this technology. Using the same measurements, we could obtain some information (limited but reasonably valid) about the gradient induced mismatch behavior for this technology. In what follows we show first how we extracted the gradient induced mismatch from the chip described in the Appendix, and afterwards we show characterization results from the second ART1 chip.

10.1 Gradient Induced Mismatch Information from Mismatch Characterization Chip

The mismatch characterization chip described in the Appendix, fabricated in a $1.0\mu m$ CMOS process (ES2 through Eurochip), contains an

8×8 array of cells. Each cell contains 30 NMOS and 30 PMOS transistors of different sizes. The size of the complete array is $2.5mm \times 2.5mm$ and distance between equal transistors is about $300\mu m$. Using this characterization chip we found out that for transistors of size $10\mu m \times 10\mu m$ spread over a chip area of $2.5mm \times 2.5mm$, biased by the same gate-to-source V_{GS} and drain-to-source V_{DS} voltages so that their nominal current was around $10\mu A$, the measured currents are as depicted in Fig-

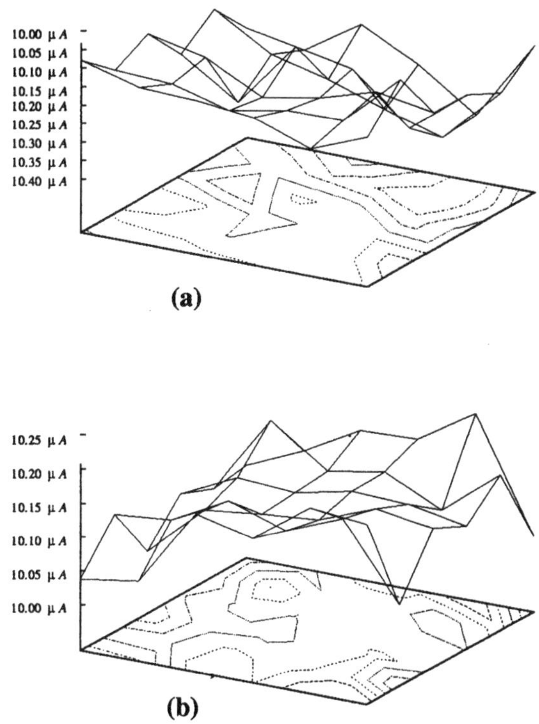

Figure 16. Measured current for an array of MOS transistors with the same V_{GS} and V_{DS} voltages (for a nominal current of $10\mu A$), spread over a die area of $2.5mm \times 2.5mm$. (a) array of NMOS transistors, (b) array of PMOS transistors.

ure 16. Figure 16(a) shows, as a function of transistor position, the current measured for each transistor of an NMOS array. Figure 16(b) shows the same for a PMOS array. As can be seen, the surfaces present a long distance gradient component and a short distance noise component. Let us call the measured currents surface $I_o(x, y)$. For this surface we can

Table 5. Current mismatch components for transistor arrays with $10\mu A$ nominal current, $10\mu m \times 10\mu m$ transistor size, and $2.5mm \times 2.5mm$ die area, for the ES2-1.0μm CMOS process.

	NMOS				PMOS			
chip	$\sigma(\Delta I_o)(\%)$	$\Delta I_o^p(\%)$	r	$\sigma_T(I_o)(\%)$	$\sigma(\Delta I_o)(\%)$	$\Delta I_o^p(\%)$	r	$\sigma_T(I_o)(\%)$
1	0.57	1.30	2.652	0.67	0.58	1.53	2.278	0.67
2	0.62	1.98	1.874	0.83	0.47	0.74	3.830	0.51
3	0.47	3.09	0.921	0.79	0.48	0.82	3.519	0.51
4	0.52	0.90	3.456	0.56	0.40	2.18	1.100	0.63
5	0.54	1.65	1.959	0.64	0.46	0.60	4.666	0.49
6	0.58	3.01	1.160	0.88	0.45	2.18	1.236	0.72
7	0.65	1.96	1.996	0.82	0.44	0.83	3.171	0.50
8	0.73	2.15	2.027	0.90	0.41	1.28	1.926	0.50

compute the best fit plane $I_o^p(x,y) = Ax + By + C$. Then, for each point (x,y) we can define

$$\Delta I_o(x,y) = I_o(x,y) - I_o^p(x,y) \tag{19}$$

By computing the standard deviation of $\Delta I_o(x,y)$, $\sigma(\Delta I_o)$, we are extracting the noise component of surface $I_o(x,y)$. The gradient component is defined by plane $I_o^p(x,y)$. The maximum deviation due to the gradient component is given by

$$\Delta I_o^p = max\left\{I_o^p(x,y)\right\} - min\left\{I_o^p(x,y)\right\} \tag{20}$$

On the other hand, for the noise component, 98% of the points remain within the $\pm 3\sigma(\Delta I_o)$ interval. Consequently, let us define the maximum deviation due to the noise component as $6\sigma(\Delta I_o)$. Let us now define

$$r = \frac{6\sigma(\Delta I_o)}{\Delta I_o^p} \tag{21}$$

as the ratio between noise component and gradient component contributions. Table 5 shows these ratios measured for NMOS and PMOS transistors of size $10\mu m \times 10\mu m$, driving nominal currents of $10\mu A$ and for different chips. Also shown in Table 5 are the standard deviations of the noise component $\sigma(\Delta I_o)$, the maximum deviation of the gradient component ΔI_o^p, and the total standard deviation of transistor currents $\sigma_T(I_o)$, computed as

$$\sigma_T(I_o) = \sqrt{\overline{I_o^2} - \overline{I_o}^2} \tag{22}$$

10.2 Details of Second Prototype Results

The current mirror tree-like structure of Figure 8 was intended to suppress the gradient component of a $1cm^2$ chip. The noise component can only be reduced by increasing transistor area [30], [31]. Table 5 reveals that for die areas of $2.5mm \times 2.5mm$, transistor sizes of $10\mu m \times 10\mu m$, and nominal currents of $10\mu A$, the contribution of noise component is equal or higher than the gradient component, while the standard deviation of current mismatch $\sigma_T(I_o)$ is kept below 1%. Consequently, for these dimensions we can avoid the use of high area consuming circuit structures (like common centroid mirrors arranged in a tree-like fashion) to eliminate the gradient component, and directly implement a single current mirror with all the outputs needed. This is the approach we used in this second ART1 prototype chip. This chip has a die area of $2.5mm \times 2.2mm$, and contains an array of 50×10 synapses, each synapse with two L_A and one L_B current sources. The current sources transistors are of size $10\mu m \times 10\mu m$ and drive a nominal current of $10\mu A$. Figure 17 shows the measured currents of the arrays. Tables 6 and 7 show the measured values of the mismatch components of the two L_A current sources arrays and Table 8 for the L_B current sources array, for all fabricated chips. Note that the total current mismatch standard deviation $\sigma_T(I_o)$ is less than 1% for all chips.

Due to the much smaller chip area its fabrication cost is much less than for the first ART1 prototype and its yield performance is significantly higher: 98% by applying eq. (18) [6]. This design implements an ART1 system with $N = 50$ input nodes and $M = 10$ category nodes, and the arrays of current sources L_A and L_B were realized directly as current mirrors of 1000 and 500 outputs, respectively, spread over an area of $2.75mm^2$. All ten fabricated chip samples were fully operational and for none of them we were able to detect any fault in its subcircuits. All system components could be isolated and independently characterized. The circuit performances of the different subcircuits were similar to those of

[6] All 10 fabricated chips were completely fault free

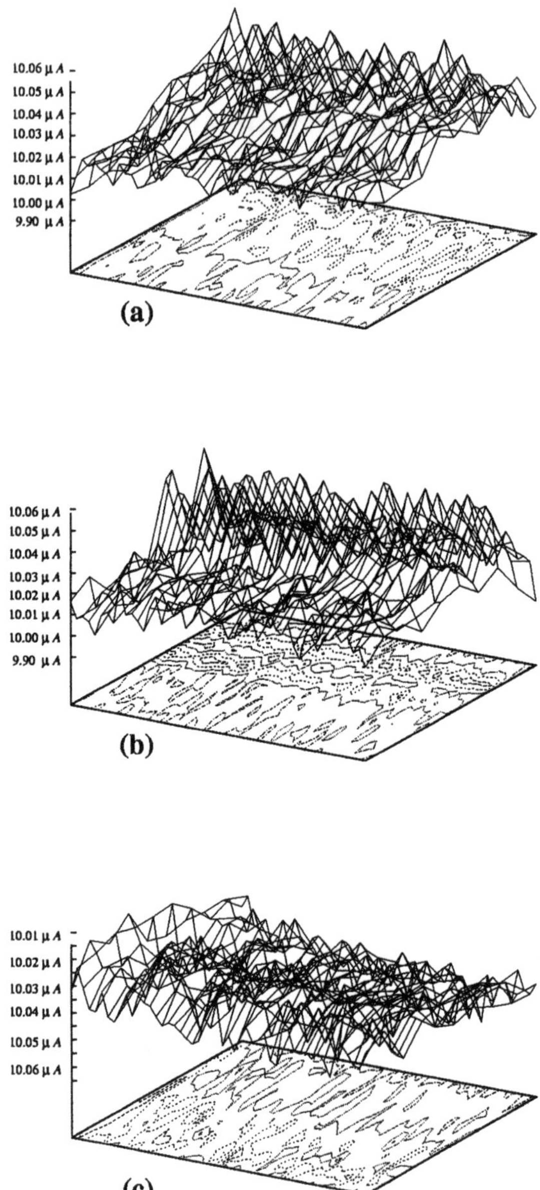

Figure 17. Measured synaptic current as a function of synapse position in the chip, for (a) L_{A1} current sources, (b) L_{A2} current sources, and (c) L_B current sources.

Table 6. Measured current error for L_{A1} NMOS current sources.

chip #	$\sigma(\Delta I_o)(\%)$	$\Delta I_o^p(\%)$	r	$\sigma_T(I_o)(\%)$
1	0.63	1.39	2.694	0.71
2	0.51	0.68	5.311	0.62
3	0.68	1.91	2.128	0.81
4	0.59	0.28	12.607	0.61
5	0.64	1.27	3.204	0.69
6	0.65	1.29	3.028	0.71
7	0.66	0.41	9.535	0.68
8	0.64	0.92	4.174	0.67
9	0.79	2.20	2.157	0.91
10	0.74	0.43	10.368	0.75

Table 7. Measured current error for L_{A2} NMOS current sources.

chip #	$\sigma(\Delta I_o)(\%)$	$\Delta I_o^p(\%)$	r	$\sigma_T(I_o)(\%)$
1	0.69	1.36	3.057	0.75
2	0.64	0.60	6.391	0.65
3	0.68	1.93	2.097	0.80
4	0.68	0.15	27.027	0.70
5	0.64	1.27	3.050	0.71
6	0.71	1.26	3.384	0.78
7	0.65	0.40	9.818	0.66
8	0.63	1.24	3.047	0.69
9	1.08	1.30	4.962	1.12
10	0.88	0.54	9.776	0.90

the first prototype, and consequently their characteristics will not be repeated. Here we will only provide some illustrative examples on system level behavior.

10.3 ART1 Systems

To test the system behavior, the chip was trained with a set of ten $7 \times 7 = 49$bit input patterns. Each pattern represents each of the ten digits from '0' to '9'. The last input pixel was always set to zero and it is not shown in the figures. The clustering test was repeated for different values of the vigilance parameter ρ and several values of parameter $\alpha = L_A/L_B$.

Table 8. Measured current error for L_B PMOS current sources.

chip #	$\sigma(\Delta I_o)(\%)$	$\Delta I_o^p(\%)$	r	$\sigma_T(I_o)(\%)$
1	0.62	0.62	6.076	0.64
2	0.59	0.22	16.497	0.60
3	0.56	3.32	1.015	0.89
4	0.63	0.90	4.196	0.64
5	0.65	1.83	2.118	0.76
6	0.64	1.49	2.565	0.73
7	0.60	1.58	2.255	0.67
8	0.62	1.48	2.524	0.71
9	0.63	0.37	10.080	0.63
10	0.57	2.16	1.573	0.73

Figure 18 shows the training sequence for $\rho = 0.3$ and $\alpha = 1.1$. The first column represents the input pattern applied to the system. The remaining ten columns correspond to the weights $\mathbf{z}_j$ stored in each category when the input pattern has been classified and learned. For each line the category with a bar to its right is the winning category after the Winner-Takes-All competition. In this case, learning self-stabilizes after two sequences of input patterns presentations. That is, no modification of the winning categories or the stored weights take place in subsequent presentations of the input pattern sequence. As shown in Figure 18, the system has clustered all ten input patterns into four categories.

A two-chip ART1 system was assembled by using the horizontal expansion approach discussed in section 7. In this case, the input patterns had $N = 100$ binary pixels. Figure 19 depicts a training sequence performed on this system. The system classifies the 10 input patterns into 8 categories after a single presentation of the input pattern set. The sequence of Figure 19 was obtained for a vigilance parameter of $\rho = 0.5$, and $\alpha = 2$ ($L_A = 10\mu A$, $L_B = 5\mu A$).

10.4 ARTMAP System

An ARTMAP hardware system can be assembled using two ART1 chips and an extra chip for the Inter-ART module, as is shown in Figure 20(a). This architecture corresponds to the "General ARTMAP" architecture [42], [13]. The Inter-ART chip, shown in Figure 20(b), is simply an ar-

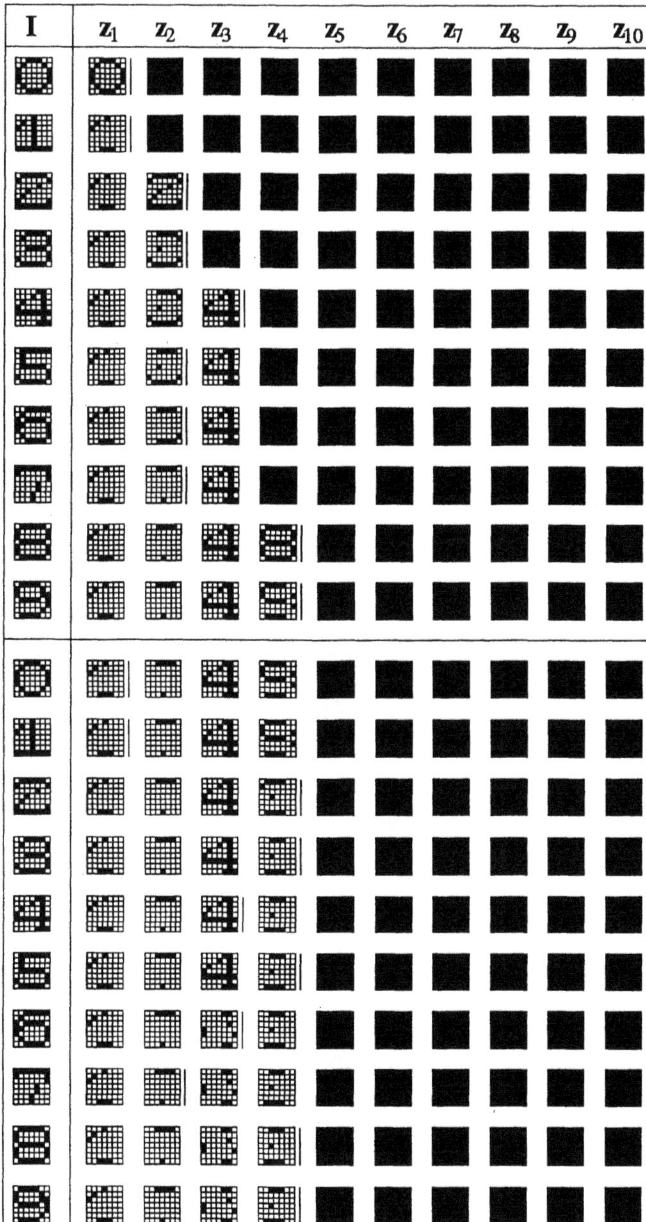

Figure 18. Training sequence for a one-chip ART1 system with $\rho = 0.3$ and $\alpha = 1.1$.

ray of cells c_{jk} whose simplified schematic is depicted in Figure 20(c). Each cell has a latch which is set initially to '1' and changes to '0' if $y_j^a = 1$, $y_k^b = 0$, and the $LEARN$ signal is high. Extra transistors, not

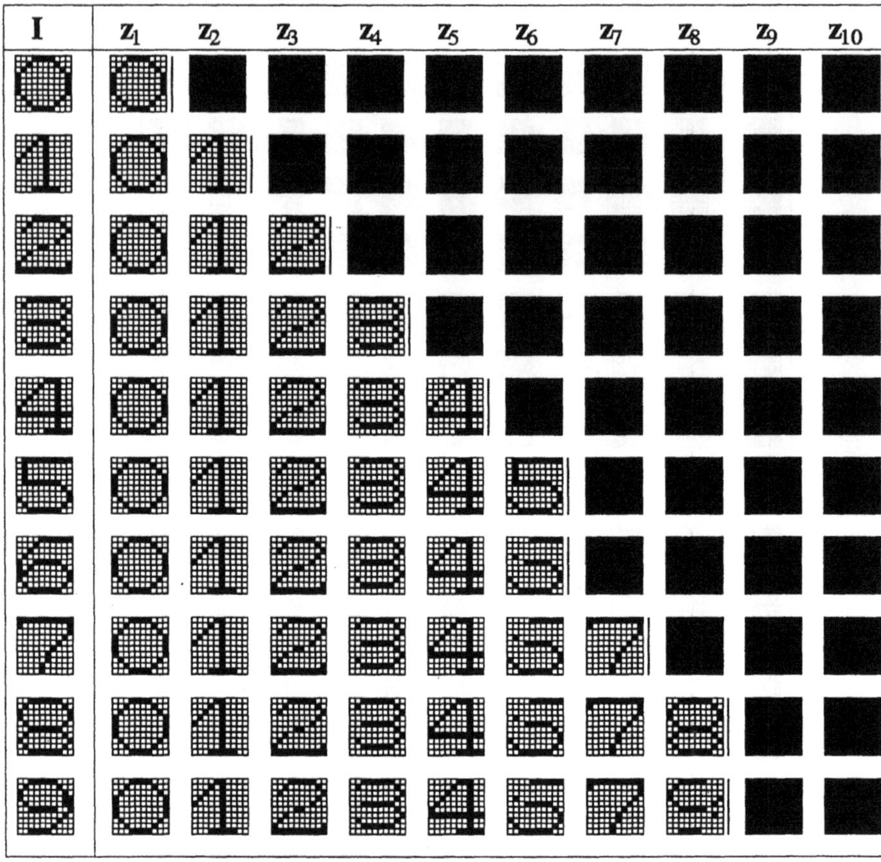

Figure 19. Training sequence for a two-chip ART1 system with $\rho = 0.5$ and $\alpha = 2$.

shown in Figure 20(c), are also included to read out the weight values. During training mode the value of weight w_{JK} is used to control a digital counter that increments the value of ρ_a. If $w_{JK} = 0$ the counter will increase its value until the ART1a winning category changes and w_{JK} becomes '1'. At this moment the counter stops and its content represents the appropriate value for ρ_a.

The system level operation of the ARTMAP hardware system has also been tested using a digital test equipment. Figure 21 shows a system training sequence. The first column, named **a**, represents the input patterns applied to the ART1a chip. The column named **b** represents the input patterns applied to the ART1b chip. The columns named $\mathbf{z}_j^a$ and $\mathbf{z}_k^b$ represent the stored weights in the ART1a and ART1b categories after

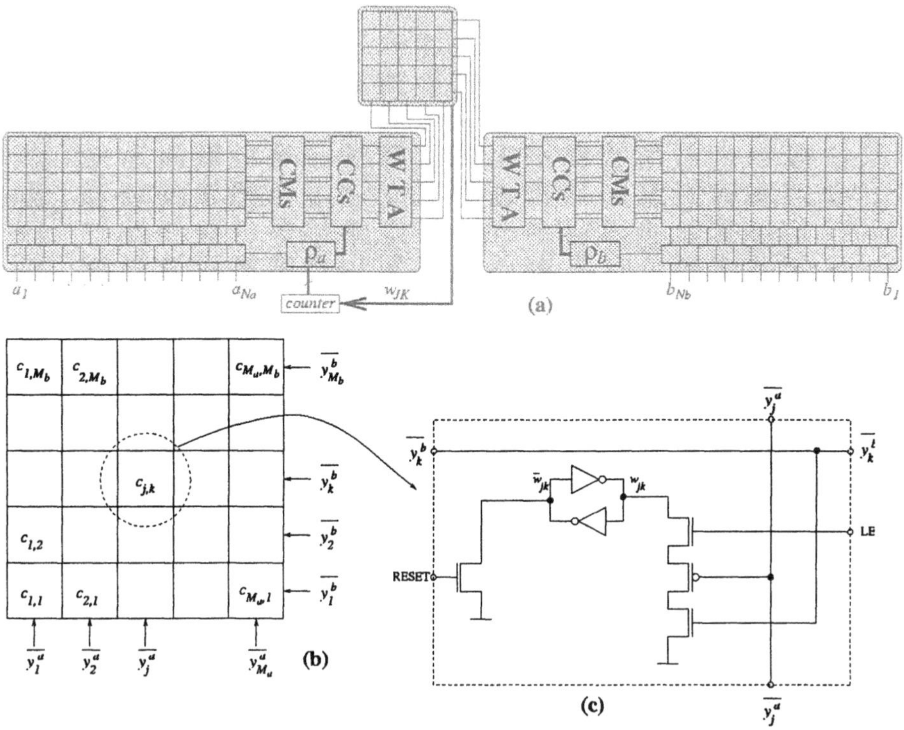

Figure 20. (a) ARTMAP hardware assembly, (b) diagram of Inter-ART chip, (c) detail of Inter-ART chip cell.

the classification and learning of each input pattern pair. The categories with a bar to their right are the ones that remain active after the search process has finished, and these are the only ones that are updated with learning. Below each ART1a winning category the final value of the vigilance parameter ρ_a needed in the search process to choose this category is indicated (ρ_a was increased in steps of $\Delta\rho_a = 1/32$). The last column shows the stored weights in the inter-ART module which represent the learned correspondence between the ART1a and ART1b categories. The vigilance parameter ρ_a was initially set to '0' and the current ratio parameter was $\alpha^a = \alpha^b = 2$ ($L_A^{a,b} = 10\mu A$ and $L_B^{a,b} = 5\mu A$). For the ART1b system it was $\rho_b = 0.75$. For this vigilance parameter, the ART1b chip forms a different category for each input pattern.

During the prediction sequence as depicted in Figure 22, only an input pattern **a** is applied to the system. The first column shows the input patterns applied to the system. These input patterns are the result of applying

Figure 21. Complete training sequence of the ARTMAP system for $\overline{\rho_a} = 0$ and $\rho_b = 0.75$.

random noise to the training patterns. The ART1a category with a bar to its right is the one chosen by this chip after the search process and the ART1b category with a bar to its right is the one the ART1a category has learned to predict through the Inter-ART weights.

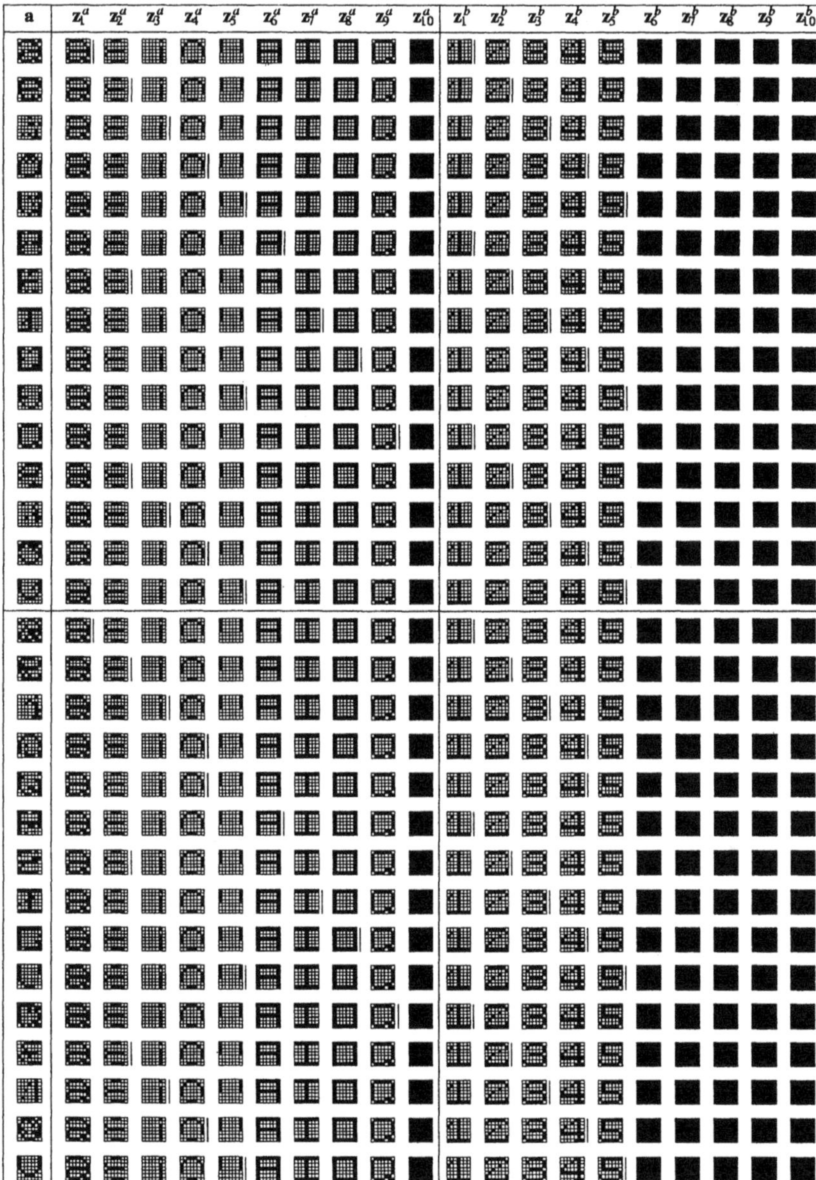

Figure 22. Recognition sequence performed on a previously trained ARTMAP system. Applied input patterns are noisy versions of the training set.

11 Appendix: CMOS Transistor Mismatch Characterization

Precise analog CMOS circuit design requires availability of confident transistor mismatch models during the design and simulation stages. During the design phase of an analog VLSI circuit, designers face many constraints imposed by the design specifications, such as speed, bandwidth, noise, precision, power consumption, area consumption, which need to be traded off for optimum overall performance. Designers must rely on accurate simulation tools in order to achieve a well optimized final design, specially if performance is pushed to the limits allowed by a given fabrication process. Simulation tools are reliable as long as they are based on good models and confident characterization techniques. If good and well characterized models are embedded in a reliable simulator, circuit designers can confidently test different circuit topologies and optimize each one of them by optimally sizing their transistors. Automatic design tools are available that by interacting with a simulator are able to obtain transistor sizes for close-to-optimum performance for a given circuit topology and a set of design constraints [43].

Many times it is not possible to simulate properly the precision limits that can be achieved by a certain circuit topology in a given fabrication process because VLSI circuit manufacturers rarely provide transistor mismatch information, and, if they do, its dependence on transistor size (width and length, independently[7]) is not known. In this Appendix we provide a very simple and cheap methodology to characterize transistor mismatch as a function of transistor width and length, and how to use this information to predict mismatch effects in circuit simulators.

In the specialized literature transistor mismatch is usually characterized by providing the standard deviation of a set of transistor electrical parameters such as the threshold voltage V_{T0}, the current gain factor $\beta = \mu C_{ox} W/L$ (μ is mobility, C_{ox} is gate oxide capacitance density, W is transistor width, and L is transistor length), and bulk threshold parameter γ. Table 9 shows a few examples [30], [33], [44], [45] on what dependencies of $\sigma^2_{(\Delta\beta/\beta)}$, $\sigma^2_{(\Delta V_{T0})}$, and $\sigma^2_{(\Delta\gamma)}$ on transistor sizes ($x = 1/W$, $y = 1/L$, W is transistor width, L is transistor length) and distance D have been postulated. A very nice study [34] based on BSIM transistor models is also available in the literature. What we present in the rest of this Chapter is a preliminary work of something that later evolved into a new mismatch model [31], but which is sufficient for the purposes de-

[7]Sometimes manufactures provide mismatch information as a function of transistor area, but this information has been obtained for (almost) square transistors [30].

scribed throughout this Chapter.

In what follows we provide an experimental method to obtain a relatively high number of samples (of $\sigma^2_{(\Delta\beta/\beta)}, \sigma^2_{(\Delta V_{TO})}, \sigma^2_{(\Delta\gamma)}$) in the $\{x, y\}$ space. Then we will fit these measured samples to a function

$$\begin{aligned}\sigma^2_{(\Delta P)} &= C_{00} + C_{10}x + C_{01}y + C_{20}x^2 + C_{11}xy + C_{02}y^2 + \cdots = \\ &= \sum_{n,m} C_{nm}x^n y^m \end{aligned} \quad (23)$$

where ΔP is the observed mismatch of a certain electrical parameter (like $\Delta\beta/\beta$, ΔV_{TO}, or $\Delta\gamma$). Note that we are not interested in discovering the physical meaning of coefficients C_{nm}, but just in obtaining a good approximation for the function $\sigma^2_{(\Delta P)} = f(x, y)$ in order to use it confidently in a circuit simulator. Note also that the $\{x, y\}$ space limits are $x_{max} = 1/W_{min}$, $y_{max} = 1/L_{min}$, $x_{min} = 0$, $y_{min} = 0$. Measuring a reasonable high number of sample points in this $\{x, y\}$ space provides sufficient information to interpolate the functions $\sigma^2_{(\Delta P)}$, which are fairly smooth in this space. Next section describes the mismatch characterization chip used to obtain all characterization data.

11.1 Mismatch Characterization Chip

According to Table 9 the mismatch in parameter P between two identical transistors is statistically characterized by a quadratic deviation whose general form is

$$\sigma^2_{(\Delta P)} = f(x, y) + S_P^2 D^2 \quad (24)$$

where $x = 1/W$, $y = 1/L$, W and L are the transistor width and length, and D is the distance between them. The two terms in eq. (24) indicate that there are two physical causes producing transistor mismatch. The first term is produced by the fact that **device physical parameters** (like doping concentrations, junctions depth, implants depth, oxide thicknesses, ...) are not exactly constant but suffer from noise-like perturbations along a die. By increasing transistor areas the **device electrical parameters** P (like threshold voltage V_{TO}, current gain factor β, or bulk

Table 9. Examples of mismatch models in the literature ($x = 1/W, y = 1/L$).

	$\sigma^2_{(\Delta\beta/\beta)}$	$\sigma^2_{(\Delta V_{TO})}$	$\sigma^2_{(\Delta\gamma)}$
[30]	$A_s^2xy + A_w^2x^2y+$ $+A_L^2xy^2 + S_\beta^2D^2$	$A^2_{V_{TO}}xy + S^2_{V_{TO}}D^2$	A_γ^2xy+ $+S_\gamma^2D^2$
[33]	$A_{\beta1}xy+$ $+A_{\beta2}(x^2+y^2)$	$A_{VO}xy$	-
[44]	A_β^2xy	$A^2_{V1}xy + A^2_{V2}xy^2-$ $-A^2_{V3}x^2y$	-

threshold parameter γ) will become less sensitive to the noisy nature of the **device physical parameters.** The second term in eq. (24), characterized by parameter S_P, is produced by the fact that the **device physical parameters** present a certain gradient variation along the dies. Usually, the gradients present in small and medium size dies can be approximated by planes. Statistical characterization of these planes (which means obtaining S_P) can be performed with a small number of transistors per die and measuring many dies. On the other hand, the transistor mismatch induced by the **device physical parameters** noisy nature, changes little from die to die. Consequently, its statistical characterization can be done by putting many transistors in a single die and measuring a reduced number of dies. This is very convenient for a 'do-it-yourself' working style, since circuit designers can easily have a small number of samples of a prototype chip at a reasonable cost. This Appendix thus concentrates on the characterization of size dependent mismatch terms, and a wide range of transistor sizes will be characterized. On the contrary, note that characterization of distance terms (like S_P) is less critical for circuit designers because gradient-induced mismatches can be compensated through layout techniques (like common centroids, for example).

With all this in mind we designed a special purpose chip intended to characterize the 'noise- induced-terms' of CMOS transistor mismatches, as a function of transistor size. As shown in Figure 23, the chip consists of an array of identical cells. Each cell contains 30 NMOS and 30 PMOS transistors, each of a different size. Sizes are such that widths are $W = 40\mu m$, $20\mu m$, $10\mu m$, $5\mu m$, $2.5\mu m$, $1.25\mu m$, and lengths are $L = 40\mu m, 10\mu m, 4\mu m, 2\mu m, 1\mu m$. Digital decoding/selection circuitry is included in each cell and outside the array. Elements in the chip are arranged in a way such that all NMOS transistors have their Drains connected to chip pin DN, all PMOS transistors have their Drains connected to chip pin DP, all NMOS and PMOS transistors have their sources connected to chip pin S, all NMOS and PMOS transistors have their Gates short-circuited to their sources, except for one NMOS-PMOS pair which has their Gates connected to chip pin G. The digital bus and the internal decoding/selection circuitry selects one cell in the array and, inside this cell, one pair of NMOS and PMOS transistors, connecting their Gates to chip pin G. A chip with an 8×8 cell array has been fabricated in a digital $1.0\mu m$ CMOS process which occupies an area of $4.0mm \times 3.5mm$, and uses 18 pins (12 for the decoding/selection Bus, DN, DP, G, S, V_{dd} , and Gnd). Some transistors in the periphery cells presented large systematic deviations with respect to those in the inside cells. Consequently, statistical computations were performed only on inner cells transistors, thus rendering an effective cell array of 6×6.

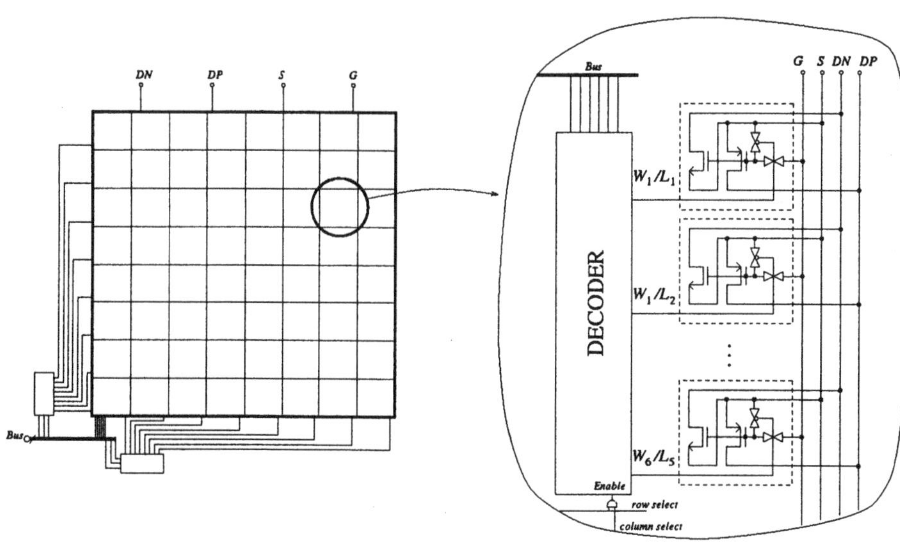

Figure 23. Mismatch characterization chip simplified schematic.

The experimental characterization set-up consists of a host computer controlling the decoding/selection bus and a DC curves measuring instrument (like the HP4145). This instrument is connected to pins DN, DP, S, G, and chip substrate. The host computer selects one NMOS-PMOS pair and the instrument measures first the NMOS transistor (putting connection DP into high-impedance and measuring through pins S, G, and DN) and then the PMOS transistor (putting connection DN into high-impedance and measuring through pins S, G, and DP). A simple software program sequentially selects and measures all transistors in the chip. Next section describes the DC curves that were measured for each transistor, how electrical parameter mismatches were extracted from these curves, and how their statistical characterization was performed.

11.2 Mismatch Parameter Extraction and Statistical Characterization

Transistor parameter mismatches were obtained by measuring pairs of identical transistors located in adjacent cells of the same rows. Since in the chip there are 6×6 effective cells, there are 6 rows, each of which provides 5 pairs of adjacent cells. This results in 30 adjacent transistor pairs (for each transistor size and type). The statistical significance of 30 measurements to determine a standard deviation is as follows: assuming a normal distribution, if 30 samples are available to compute a standard deviation $\sigma_{Computed}$, it can be assured that the 95% confidence interval for the real standard deviation σ_{Real} is [46]

$$0.7964 \times \sigma_{Computed} \leq \sigma_{Real} \leq 1.344 \times \sigma_{Computed}. \tag{25}$$

For each transistor pair, two curves were measured while operating in the ohmic region (always in strong inversion). These curves are[8]

$$\textbf{Curve 1:} \quad I_{DS}(V_{GS}), V_{SB} = 0V, \quad V_{DS} = 0.1V, V_{GS} \in [V_{GS_{min}}, V_{GS_{max}}] \tag{26}$$

$$\textbf{Curve 2:} \quad I_{DS}(V_{SB}), V_{GS} = 3.0V, \quad V_{DS} = 0.1V, V_{SB} \in [V_{SB_{min}}, V_{SB_{max}}] \tag{27}$$

Care must be taken in order to keep current levels sufficiently small so that mismatch introduced by series resistances (contact resistances, vari-

[8] All voltages and currents are taken in absolute value, so that the same expressions are valid for NMOS and PMOS transistors.

able length routing wires, ...) is negligible. The following strong inversion ohmic region transistor model was assumed,

$$I_{DS} = \beta \frac{V_{GS} - V_T(V_{SB}) - \frac{1}{2}V_{DS}}{1 + \theta(V_{GS} - V_T(V_{SB}))} V_{DS} \tag{28}$$

$$V_T(V_{SB}) = V_{T0} + \gamma \left[\sqrt{\Phi + V_{SB}} - \sqrt{\Phi} \right] \tag{29}$$

which renders the following current mismatch for each transistor pair:

$$\begin{aligned} \frac{\Delta I_{DS}}{I_{DS}} &= \frac{\Delta\beta}{\beta} - \frac{1 + \frac{1}{2}\theta V_{DS}}{V_{GS} - V_T(V_{SB}) - \frac{1}{2}V_{DS}} \Delta V_T - \\ &- \frac{V_{GS} - V_T(V_{SB})}{1 + \theta(V_{GS} - V_T(V_{SB}))} \Delta\theta \end{aligned} \tag{30}$$

$$\Delta V_T = \Delta V_{T0} + \Delta\gamma \left[\sqrt{\Phi + V_{SB}} - \sqrt{\Phi} \right] \tag{31}$$

The drain and source series resistances (due to contacts, diffusion resistance, and metal routing lines) have the effect of changing the extracted value of the mobility reduction parameter θ in eq. (28) [30], [31],

$$\theta = \theta_{Real} + \beta R_{DS} \tag{32}$$

where θ_{Real} is the real mobility reduction parameter of the transistor and R_{DS} is the sum of the series resistances at drain and source. The mismatch contribution of $\Delta(\beta R_{DS})$ can be of the order or higher than that of $\Delta\theta_{Real}$, but both contribute very little to $\Delta I_{DS}/I_{DS}$.

For each transistor pair, the measurement/extraction procedure was as follows:

- **Curve 1** (eq. (27)) was measured for both transistors. Using the Levenberg-Marquardt nonlinear curve fitting technique, the first curve was fitted to eq. (28) and parameters β, V_{T0}, and θ were obtained. Using the two measured curves, the curve $\Delta I_{DS}/I_{DS}$ was computed and fitted to eq. (31) obtaining $\Delta\beta/\beta$, ΔV_{T0}, and $\Delta\theta$ for this transistor pair.

- **Curve 2** (eq. (28)) was measured for both transistors, and curve $\Delta I_{DS}/I_{DS}$ was computed. According to eqs. (31), (31) it must fit to

$$\frac{\Delta I_{DS}}{I_{DS}} = \qquad (33)$$
$$\frac{\Delta\beta}{\beta} - \frac{1+\frac{1}{2}\theta V_{DS}}{V_x - \frac{1}{2}V_{DS}}\left[\Delta V_{T0} + \Delta\gamma(\sqrt{\Phi + V_{SB}} - \sqrt{\Phi})\right] -$$
$$-\frac{V_x}{1+\theta V_x}\Delta\theta,$$

where $V_x = V_{GS} - V_T(V_{SB})$. For this transistor pair $\Delta\beta/\beta$ and ΔV_{T0} are already known. The values for $V_x = V_{GS} - V_T(V_{SB})$ can be obtained from eq. (28)

$$V_x = \frac{I_{DS} + \frac{\beta}{2}V_{DS}^2}{\beta V_{DS} - \theta I_{DS}} \qquad (34)$$

since β, θ, and V_{DS} are already known, and the I_{DS} values are those just measured at Curve 2. Consequently, $\Delta\gamma$ is the only unknown parameter in eq. (34), which can be obtained for this pair by fitting eq. (34), after substituting eq. (34) into it.

This measurement/extraction procedure is repeated for the $N_T = 30$ transistor pairs. For each extracted mismatch parameter ΔP ($\Delta\beta/\beta$, ΔV_{T0}, $\Delta\gamma$) its standard deviation

$$\sigma^2_{(\Delta P)} = \frac{1}{N_T}\sum_{n=1}^{N_T}(\Delta P_n - \overline{\Delta P})^2 \qquad (35)$$

is computed. For each fabricated chip, eq. (35) should be obtained for each transistor size and type (NMOS or PMOS).

11.3 Characterization Results

A mismatch characterization chip was fabricated in a digital double-metal single-poly $1.0\mu m$ CMOS process. The die area of the chip is $3.5mm \times 4.0mm$. Ten samples were delivered by the foundry, eight of which were fault free. For each die, transistor size, and transistor type the following parameters were extracted, following the procedure described in the previous section,

$$\sigma_{(\Delta\beta/\beta)} \quad , \quad \sigma_{(\Delta V_{TO})} \quad , \quad \sigma_{(\Delta\gamma)} \tag{36}$$

Table 10 shows these parameters for the NMOS transistors, averaged over all dies and indicating the spread from die to die. Table 11 shows the average over all dies of these parameters for the PMOS transistors. Each cell in Tables 10 and 11 indicates

$$\overline{\sigma_{(\Delta P)}} \pm 3\sigma(\sigma_{(\Delta P)}) \tag{37}$$

where,

$$\overline{\sigma_{(\Delta P)}} = \frac{1}{N_{Dies}} \sum_{n_d=1}^{N_{Dies}} \sigma_{(\Delta P)}(n_d) \tag{38}$$

$$\sigma^2(\sigma_{(\Delta P)}) = \frac{1}{N_{Dies}} \sum_{n_d=1}^{N_{Dies}} \left(\sigma_{(\Delta P)}(n_d) - \overline{\sigma_{(\Delta P)}}\right)^2$$

and N_{Dies} is the total number of fault-free dies.

Looking at the deviations $\sigma_{(\Delta P)}$ in Tables 10 and 11 and at their $\pm 3\sigma$ inter-chip spread, one can see that the 3σ spread is of the order of $\pm 50\%$ of the average deviation. This shows that the deviations $\sigma_{(\Delta P)}$ of transistor mismatch parameters have a fairly stable behavior from chip to chip.

In order to measure the precision of the extracted deviation values, one of the dies was measured without sweeping the transistor pairs: Curves 1 and 2 (eqs. (27) and (28)) were measured 30 times for the same pair. The resulting deviations indicate the measurement set-up and extraction procedure precision. This is given in Table 12 for the NMOS transistors and in Table 13 for the PMOS transistors. Units in Table 12 and 13 are the same than for Tables 10 and 11, except for a 10^{-3} factor.

The extracted data for each column in Tables 10 and 11 can be considered to be sample points of a two dimensional surface whose independent variables are $x = 1/W$ and $y = 1/L$,

$$\sigma_{(\Delta P)} = f(x, y) \tag{39}$$

Table 10. NMOS mismatch characterization parameters averaged over all measured dies and indicating the $\pm 3\sigma$ spread from die to die.

sizes	$\sigma_{(\Delta\beta/\beta)}(\times 10^{-3})$	$\sigma_{(\Delta V_{T0})}(mV)$	$\sigma_{(\Delta\gamma)}(mV^{1/2})$
40/40	1.18 ± 0.78	0.60 ± 0.35	0.49 ± 0.15
40/10	2.21 ± 1.71	1.21 ± 0.59	0.57 ± 0.22
40/4	3.36 ± 1.71	1.27 ± 0.52	0.86 ± 0.36
40/2	5.92 ± 1.41	1.94 ± 0.76	1.29 ± 0.71
40/1	9.01 ± 2.93	5.23 ± 1.97	3.93 ± 1.28
20/40	1.87 ± 0.66	0.66 ± 0.25	0.48 ± 0.27
20/10	2.43 ± 0.90	1.14 ± 0.40	0.73 ± 0.34
20/4	4.17 ± 1.75	1.67 ± 0.56	1.14 ± 0.46
20/2	9.18 ± 4.74	2.58 ± 1.84	1.80 ± 1.07
20/1	12.52 ± 7.10	6.67 ± 1.95	4.72 ± 1.19
10/40	1.77 ± 0.51	0.66 ± 0.24	0.62 ± 0.49
10/10	4.87 ± 1.45	1.37 ± 0.41	0.92 ± 0.43
10/4	6.83 ± 1.94	2.03 ± 0.88	1.39 ± 0.37
10/2	7.89 ± 2.23	3.70 ± 1.69	2.27 ± 1.29
10/1	13.41 ± 7.36	8.60 ± 2.30	5.49 ± 3.23
5/40	2.97 ± 1.26	0.82 ± 0.61	0.76 ± 0.25
5/10	4.53 ± 2.28	2.01 ± 0.68	1.35 ± 0.56
5/4	6.77 ± 1.90	3.29 ± 1.04	2.02 ± 1.24
5/2	10.10 ± 4.28	4.76 ± 2.21	3.13 ± 1.14
5/1	14.80 ± 4.00	11.66 ± 3.10	6.15 ± 2.54
2.5/40	6.71 ± 2.88	1.27 ± 0.86	1.01 ± 0.40
2.5/10	8.84 ± 5.01	2.46 ± 1.26	1.66 ± 0.78
2.5/4	9.09 ± 3.69	4.14 ± 1.43	2.98 ± 1.50
2.5/2	13.85 ± 5.46	6.68 ± 3.32	4.30 ± 1.05
2.5/1	21.56 ± 10.77	15.93 ± 8.49	8.51 ± 3.28
1.25/40	11.40 ± 4.56	1.67 ± 0.87	1.79 ± 0.56
1.25/10	12.50 ± 4.96	3.52 ± 1.92	2.69 ± 1.34
1.25/4	10.71 ± 2.58	5.24 ± 2.06	3.75 ± 1.53
1.25/2	15.26 ± 8.25	9.65 ± 4.65	5.69 ± 2.66
1.25/1	21.69 ± 9.06	21.94 ± 6.02	11.06 ± 4.29

Table 11. PMOS mismatch characterization parameters averaged over all measured dies and indicating the $\pm 3\sigma$ spread from die to die.

sizes	$\sigma_{(\Delta\beta/\beta)}(\times 10^{-3})$	$\sigma_{(\Delta V_{T0})}(mV)$	$\sigma_{(\Delta\gamma)}(mV^{1/2})$
40/40	1.41 ± 0.67	1.38 ± 0.77	0.45 ± 0.18
40/10	2.20 ± 1.35	1.89 ± 1.54	0.53 ± 0.20
40/4	3.10 ± 1.41	1.87 ± 0.73	0.64 ± 0.16
40/2	6.22 ± 2.70	2.73 ± 0.69	1.08 ± 0.41
40/1	11.43 ± 6.39	5.78 ± 2.33	3.65 ± 1.70
20/40	1.75 ± 0.49	1.06 ± 0.87	0.54 ± 0.20
20/10	2.97 ± 1.20	1.80 ± 0.70	0.58 ± 0.19
20/4	3.77 ± 1.56	2.27 ± 1.09	0.83 ± 0.21
20/2	10.06 ± 4.41	3.56 ± 0.80	1.59 ± 0.89
20/1	13.55 ± 4.92	7.60 ± 2.61	3.81 ± 1.83
10/40	1.64 ± 0.74	1.19 ± 0.50	0.49 ± 0.22
10/10	4.76 ± 2.06	2.18 ± 1.10	0.73 ± 0.32
10/4	6.75 ± 2.03	2.78 ± 0.72	1.00 ± 0.52
10/2	9.03 ± 7.84	4.86 ± 3.14	1.93 ± 0.78
10/1	20.70 ± 6.41	10.66 ± 2.77	5.18 ± 1.83
5/40	2.89 ± 1.44	1.43 ± 0.56	0.55 ± 0.30
5/10	5.21 ± 1.60	2.54 ± 0.78	0.91 ± 0.51
5/4	6.86 ± 1.67	4.33 ± 1.90	1.35 ± 0.47
5/2	11.23 ± 7.76	7.56 ± 4.29	2.22 ± 0.52
5/1	17.99 ± 8.99	12.81 ± 4.41	4.98 ± 2.17
2.5/40	6.04 ± 1.94	1.73 ± 0.69	0.58 ± 0.23
2.5/10	9.44 ± 6.28	3.40 ± 0.77	1.12 ± 0.50
2.5/4	11.82 ± 4.01	5.86 ± 1.37	1.94 ± 0.82
2.5/2	14.39 ± 8.64	9.08 ± 4.18	3.14 ± 1.54
2.5/1	25.44 ± 8.31	17.63 ± 5.36	6.60 ± 2.40
1.25/40	12.22 ± 8.05	2.65 ± 1.73	0.95 ± 0.40
1.25/10	14.32 ± 7.29	4.50 ± 2.46	1.71 ± 0.70
1.25/4	12.46 ± 5.98	6.70 ± 3.00	2.49 ± 0.94
1.25/2	19.33 ± 10.56	10.76 ± 4.12	3.80 ± 1.49
1.25/1	28.54 ± 9.41	21.05 ± 7.97	7.18 ± 3.89

Table 12. NMOS precision measurements for the data in Table 10.

sizes	$\sigma_{(\Delta\beta/\beta)}$	$\sigma_{(\Delta V_{T0})}(V)$	$\sigma_{(\Delta\gamma)}(V^{1/2})$
40/40	2.99e-04	1.70e-04	2.64e-04
40/10	4.29e-04	2.54e-04	3.58e-04
40/4	3.84e-04	2.09e-04	4.36e-04
40/2	4.57e-04	2.88e-04	4.69e-04
40/1	6.50e-04	4.15e-04	6.62e-04
20/40	2.64e-04	1.40e-04	3.14e-04
20/10	2.80e-04	1.59e-04	3.29e-04
20/4	3.56e-04	1.92e-04	3.07e-04
20/2	4.35e-04	2.24e-04	4.98e-04
20/1	5.58e-04	3.92e-04	6.02e-04
10/40	3.92e-04	2.46e-04	3.10e-04
10/10	3.87e-04	2.39e-04	4.17e-04
10/4	4.51e-04	3.04e-04	4.00e-04
10/2	4.20e-04	2.44e-04	3.53e-04
10/1	6.11e-04	4.20e-04	7.70e-04
5/40	3.75e-04	1.71e-04	4.45e-04
5/10	4.38e-04	2.57e-04	3.71e-04
5/4	2.91e-04	1.80e-04	2.94e-04
5/2	3.49e-04	2.53e-04	3.76e-04
5/1	4.60e-04	2.47e-04	7.99e-04
2.5/40	2.80e-04	1.25e-04	3.68e-04
2.5/10	3.49e-04	2.32e-04	4.20e-04
2.5/4	3.64e-04	2.13e-04	4.77e-04
2.5/2	3.52e-04	2.07e-04	4.06e-04
2.5/1	4.20e-04	3.29e-04	4.97e-04
1.25/40	2.72e-04	1.76e-04	2.60e-04
1.25/10	3.54e-04	1.69e-04	3.24e-04
1.25/4	3.68e-04	2.27e-04	3.24e-04
1.25/2	4.20e-04	2.70e-04	4.53e-04
1.25/1	4.86e-04	2.55e-04	4.63e-04

Table 13. PMOS precision measurements for the data in Table 11.

sizes	$\sigma_{(\Delta\beta/\beta)}$	$\sigma_{(\Delta V_{T0})}(V)$	$\sigma_{(\Delta\gamma)}(V^{1/2})$
40/40	4.99e-04	2.95e-04	4.10e-04
40/10	4.01e-04	1.77e-04	4.70e-04
40/4	3.70e-04	1.60e-04	3.73e-04
40/2	4.23e-04	1.56e-04	4.06e-04
40/1	6.46e-04	2.67e-04	6.66e-04
20/40	4.67e-04	1.89e-04	3.70e-04
20/10	4.57e-04	2.24e-04	3.74e-04
20/4	3.77e-04	1.21e-04	4.60e-04
20/2	3.70e-04	1.98e-04	3.83e-04
20/1	5.53e-04	2.53e-04	4.98e-04
10/40	4.80e-04	1.97e-04	3.78e-04
10/10	4.22e-04	2.37e-04	3.86e-04
10/4	4.64e-04	1.80e-04	4.03e-04
10/2	5.29e-04	2.01e-04	4.74e-04
10/1	4.16e-04	2.33e-04	3.71e-04
5/40	3.43e-04	1.75e-04	3.51e-04
5/10	5.26e-04	1.74e-04	4.38e-04
5/4	4.02e-04	2.32e-04	3.76e-04
5/2	4.21e-04	2.05e-04	3.57e-04
5/1	4.85e-04	2.64e-04	4.81e-04
2.5/40	4.52e-04	2.81e-04	3.45e-04
2.5/10	4.43e-04	1.60e-04	3.91e-04
2.5/4	3.63e-04	1.26e-04	3.22e-04
2.5/2	3.40e-04	1.06e-04	4.04e-04
2.5/1	4.02e-04	2.37e-04	5.01e-04
1.25/40	4.26e-04	2.50e-04	5.03e-04
1.25/10	2.05e-04	6.80e-05	2.76e-04
1.25/4	3.34e-04	1.08e-04	3.08e-04
1.25/2	4.84e-04	1.41e-04	4.87e-04
1.25/1	6.41e-04	3.23e-04	5.65e-04

As mentioned before, several functionals have been attributed to eq. (39). We will assume the following polynomial dependency

$$\sigma^2_{(\Delta P)_{fit}}(x, y, C_{nm}) =$$
$$C_{00}+C_{10}x+C_{01}y+C_{20}x^2+C_{11}xy+C_{02}y^2+\cdots == \sum_{n,m} C_{nm}x^n y^m \quad (40)$$

The optimum set of coefficients C_{nm} were obtained by fitting eq. (40) (using Least Mean Squares) to the experimental data $\sigma_{(\Delta P)}(x_i, y_i, n_d)$ for all sizes and dies. For this, the following error term was defined,

$$\epsilon = \sum_{n_d=1}^{N_{Dies}} \sum_{i=1}^{N_{Sizes}} w_i \left[\sigma^2_{(\Delta P)}(x_i, y_i, n_d) - \sigma^2_{(\Delta P)_{fit}}(x_i, y_i, C_{nm})\right]^2 \quad (41)$$

where w_i is a weighting term that depends on the spread of $\sigma_{(\Delta P)}(x_i, y_i, n_d)$ from die to die, and N_{Sizes} is the total number of transistor sizes available in the chip. If $\sigma_{(\Delta P)}$, for size $x_i = 1/W_i$, $y_i = 1/L_i$ has a large spread from die to die then weight w_i will be smaller than for another size whose spread is smaller. If for a given size (x_i, y_i) the spread from die to die is $\sigma(\sigma_{(\Delta P)}(x_i, y_i))$ then the weight w_i is defined as

$$w_i = \frac{e^{-\Omega_i}}{\Omega_i^2} \quad , \quad \Omega_i = \frac{\sigma(\sigma_{(\Delta P)}(x_i, y_i))}{\sigma_{(\Delta P)}(x_i, y_i)}. \quad (42)$$

Figure 24(a) shows the resulting fitted surface for $\sigma_{(\Delta\beta/\beta)_{fit}}(1/W, 1/L)$, for NMOS transistors. Also shown in Figure 24(a) (with diamonds) are the experimental measured values for $\sigma_{(\Delta\beta/\beta)}(1/W_i, 1/L_i, n_d)$ for each transistor size and for each die, for NMOS transistors. Figure 24(b) shows the fitted surface and experimental data for the threshold voltage deviation $\sigma_{(\Delta V_{TO})}$, and Figure 24(c) does it for the bulk threshold parameter deviation $\sigma_{(\Delta\gamma)}$, for NMOS transistors. The coefficients C_{nm} that result from the fitting procedure are given in Tables 14 and 15 for all deviations for NMOS and PMOS transistors, respectively. The units

Table 14. NMOS resulting coefficients for the fitting functions.

	C_{00}	C_{11}	C_{20}	C_{02}	C_{21}	C_{12}	C_{22}
$\sigma^2_{(\Delta\beta/\beta)}$	1.7e-06	8.9e-04	2.0e-04	5.1e-05	-1.2e-03	4.2e-04	0
$\sigma^2_{(\Delta V_{TO})}$	3.4e-07	-4.3e-05	8.1e-06	1.6e-05	0	6.2e-04	0
$\sigma^2_{(\Delta\gamma)}$	2.0e-07	-7.5e-06	5.1e-06	1.3e-05	3.3e-05	1.6e-04	-5.7e-05

Table 15. PMOS resulting coefficients for the fitting functions.

	C_{00}	C_{11}	C_{20}	C_{02}	C_{21}	C_{12}	C_{22}
$\sigma^2_{(\Delta\beta/\beta)}$	1.4e-06	7.1e-04	2.6e-04	1.5e-04	-6.0e-04	9.1e-04	-6.5e-04
$\sigma^2_{(\Delta V_{TO})}$	2.5e-06	1.7e-04	0	0	-1.3e-04	8.4e-04	-4.5e-04
$\sigma^2_{(\Delta\gamma)}$	1.7e-07	-2.0e-05	0	1.0e-05	5.5e-05	1.3e-04	-1.3e-04

of the coefficients are such that if x and y are expressed in μm^{-1} then the deviations σ have the units used in Tables 10, 11 (without the 10^{-3} factor), and Tables 12, 13.

Correlations among deviations of different parameters can also be easily obtained. However, according to our data, no definite conclusions can be made. We observed that for some dies there were high correlation coefficients for some parameter pairs, while for other dies the same correlation coefficient could change significantly (we observed changes from almost $+1$ correlation factor for one die to almost -1 for another die). If these measurements are correct this would mean that deviation correlations could change significantly from one area of the wafer to another. However, before making any conclusions about this possible "wafer-level" behavior it would be necessary to make intensive measurements using many dies per wafer and for many wafers. In our case, since we only used 8 dies from one run we can only make approximate conclusions regarding the measured standard deviation values, but not much can be said about their correlations.

12 Conclusions

ART1 algorithms are very powerful for Real-Time clustering applications. With a minor modification they have many VLSI-attractive features. We have provided a circuit design technique to map the full func-

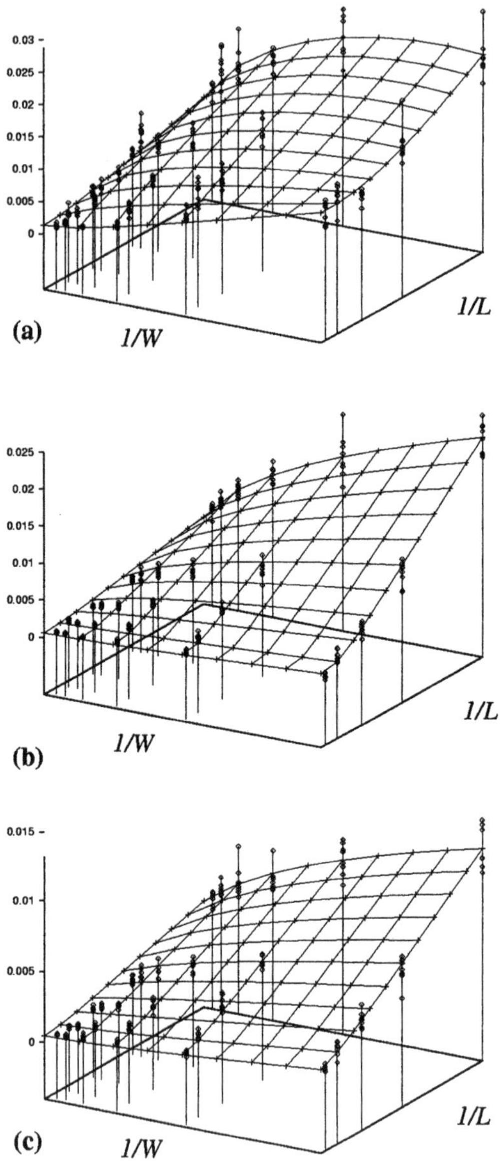

Figure 24. Experimentally measured/extracted mismatch data (diamonds) as a function of transistor size, for NMOS transistors. Also shown are the interpolated surfaces. (a) Results for $\sigma_{(\Delta\beta/\beta)}$, (b) for $\sigma_{(\Delta V_{T0})}$, (c) and for $\sigma_{(\Delta\gamma)}$.

tionality of the ART1 algorithm into a highly efficient hardware microchip. Experimental results on single- and multichip ART1 systems are provided.

References

[1] Mauduit, N., Duranton, M., Gobert, J., and Sirat, J.A. (1992), "Lneuro 1.0: A piece of hardware lego for building neural network systems," *IEEE Transactions on Neural Networks*, Vol. 3, no. 3, pp. 414-422.

[2] Jones, S., Sammut, K., Nielsen, C., and J. Staunstrup, (1991), "Toroidal neural network: Architecture and processor granularity issues," in *VLSI Design of Neural Networks*, U. Ramacher and U. Rueckert, Eds., pp. 229-254. Kluwer Academic Publishers, Dordrecht, Netherlands.

[3] Ramacher, U., Beichter, J., Raab, W., Anlauf, J., Bruels, N., Hachmann, U., and Wesseling, M. (1991), "Design of a 1st generation neurocomputer," in *VLSI Design of Neural Networks*, Ramacher, U. and Rueckert, U. Eds., pp. 271-310. Kluwer Academic Publishers, Dordrecht, Netherlands.

[4] Kohonen, T. (1989), *Self-Organization and Associative Memory*, Springer-Verlag, Berlin, Germany.

[5] Bezdek, J. (1981), *Pattern Recognition with Fuzzy Objective Function Algorithms*, Plenum, New York.

[6] Duda, R. and Hart, P. (1973), *Pattern Classification and Scene Analysis*, Wiley, New York.

[7] Hartigan, J. (1975), *Clustering Algorithms*, Wiley, New York.

[8] Dubes, R. and Jain, A. (1988), *Algorithms that Cluster Data*, Prentice Hall, Englewood Cliffs, NJ.

[9] Pao, Y.H. (1989), *Adaptive Recognition and Neural Networks*, Addison Wesley, Reading, MA.

[10] Rodríguez-Vázquez, A., Espejo, S., Domínguez-Castro, R., Huertas, J.L., and Sánchez-Sinencio, E. (1993), "Current-mode techniques for the implementation of continous- and discrete-time cellular neural networks," *IEEE Transactions on Circuits and Systems II*, Vol. 40, no. 3, pp. 132-146.

[11] Linares-Barranco, B., Sánchez-Sinencio, E., Rodríguez-Vázquez, A., and Huertas, J.L. (1993), "A CMOS analog adaptive BAM with on-chip learning and weight refreshing," *IEEE Transactions on Neural Networks*, Vol. 4, no. 3, pp. 445-455.

[12] Carpenter, G.A. and Grossberg, S. (1987), "A massively parallel architecture for a self-organizing neural pattern recognition machine," *Computer Vision, Graphics, and Image Processing*, Vol. 37, pp. 54-115.

[13] Serrano-Gotarredona, T. and Linares-Barranco, B. (1998), *Adaptive Resonance Theory Microchips. Circuit Design Techniques*, Kluwer Academic Publishers, Boston.

[14] Keulen, E., Colak, S., Withagen, H., and Hegt, H. (1994), "Neural network hardware performance criteria," *Proceedings of the 1994 Int. Conf. on Neural Networks (ICNN'94)*, pp. 1885-1888.

[15] Riedmiller, M. (1994), "Advanced supervised learning in multilayer perceptrons - from backpropagation to adaptive learning algorithms," *Int. Journal of Computer Standards and Interfaces*, Vol. 5.

[16] Kosko, B. (1987), "Adaptive bidirectional associative memories," *Applied Optics*, Vol. 26, pp. 4947-4960.

[17] Wand, Y.F., Cruz, J.B., and Mulligan, J.H. (1990), "On multiple training for bidirectional associative memory," *IEEE Transactions on Neural Networks*, Vol. 1, pp. 275-276.

[18] Linares-Barranco, B., Sánchez-Sinencio, E., Rodríguez-Vázquez, A., and Huertas, J.L. (1992), "A modular T-mode design approach for analog neural network hardware implementations," *IEEE Journal of Solid-State Circuits*, Vol. 27, no. 5, pp. 701-713.

[19] Ho, C.S., Liou, J.J., Georgiopoulos, M., Heileman, G.L., and Christodoulo, C. (1994), "Analogue circuit design and implementation of an adaptive resonance theory (ART) neural network architecture," *International Journal of Electronics*, Vol. 76, no. 2, pp. 271-291.

[20] Tsay, S.W. and Newcomb, R.W. (1991), "VLSI implementation of ART1 memories," *IEEE Transactions on Neural Networks*, Vol. 2, pp. 214-221, March.

[21] Wunsch II, D.C., Caudell, T.P., Capps, C.D., Marks II, R.J., and Falk, R.A. (1993), "An optoelectronic implementation of the adaptive resonance neural network," *IEEE Transactions on Neural Networks*, Vol. 4, no. 4, pp. 673-684, July.

[22] Serrano-Gotarredona, T. and Linares-Barranco, B. (1996), "A modified ART1 algorithm more suitable for VLSI implementations," *Neural Networks*, Vol. 9, no. 6, pp. 1025-1043.

[23] Serrano-Gotarredona, T. and Linares-Barranco, B. (1996), "A real-time clustering microchip neural engine," *IEEE Transactions on Very Large Scale Integration (VLSI) Systems*, Vol. 4, no. 2, pp. 195-209.

[24] Lazzaro, J., Ryckebush, R., Mahowald, M.A., and Mead, C. (1989), "Winner-take-all networks of O(n) complexity," in *Advances in Neural Information Processing Systems*, Touretzky, D.S. Ed., Vol. 1, pp. 703-711. Morgan-Kaufmann, Los Altos, CA.

[25] Rodríguez-Vázquez, A., Domínguez-Castro, R., Medeiro, F., and Delgado-Resti-tuto, M. (1995), "High resolution CMOS current comparators: Design and application to current-mode function generation," *International Journal of Analog Integrated Circuits and Signal Processing*, Vol. 7, pp. 149-165.

[26] Mead, C. (1989), *Analog VLSI and Neural Systems*, Addison Wesley, Reading, MA.

[27] Serrano-Gotarredona, T. and Linares-Barranco, B. (1994), "The active-input regulated-cascode current mirror," *IEEE Transactions*

on Circuits and Systems, Part I: Fundamental Theory and Applications, Vol. 41, no. 6, pp. 464-467.

[28] Serrano-Gotarredona, T. and Linares-Barranco, B. (1995), "A modular current-mode high-precision winner-take-all circuit," *IEEE Transactions on Circuits and Systems, Part II*, Vol. 42, no. 2, pp. 132-134.

[29] Serrano-Gotarredona, T. and Linares-Barranco, B. (1998), "A high-precision current-mode WTA-MAX circuit with multi-chip capability," *IEEE Journal of Solid-State Circuits*, Vol. 33, no. 2, pp. 280-286.

[30] Pelgrom, M.J.M., Duinmaijer, A.C.J., and Welbers, A.P.G. (1989), "Matching properties of MOS transistors," *IEEE Journal of Solid-State Circuits*, Vol. 24, pp. 1433-1440.

[31] Serrano-Gotarredona, T. and Linares-Barranco, B. (1999), "Systematic width-and-length dependent cmos transistor mismatch characterization and simulation," *International Journal of Analog Integrated Circuits and Signal Processing*.

[32] Gregor, R.W. (1992), "On the relationship between topography and transistor matching in an analog CMOS technology," *IEEE Transactions on Electron Devices*, Vol. 39, no. 2, pp. 275-282.

[33] Lakshmikumar, K.R., Hadaway, R.A., and Copeland, M.A. (1986), "Characterization and modeling of mismatch in MOS transistors for precision analog design," *IEEE Journal of Solid-State Circuits*, Vol. SC-21, no. 6, pp. 1057-1066.

[34] Michael, C. and Ismail, M. (1992), "Statistical modeling of device mismatch for analog MOS integrated circuits," *IEEE Journal of Solid-State Circuits*, Vol. 27, no. 2, pp. 154-166.

[35] Shyu, J.B., Temes, G.C., and Krummenacher, F. (1984), "Random error effects in matched MOS capacitors and current sources," *IEEE Journal of Solid-State Circuits*, Vol. SC-19, no. 6, pp. 948-955.

[36] Allen, P.E. and Holberg, D.R. (1987), *CMOS Analog Design*, Holt Rinehart and Winston Inc., New York.

[37] Strader, N.R. and Harden, J.C. (1989), "Architectural yield optimization," in *Wafer Scale Integration*, Schwartzlander, E.E. Ed., pp. 57-118. Kluwer Academic Publishers, Boston.

[38] Chu, L.C. (1991), "Fault-tolerant model of neural computing," *IEEE Int. Conf. on Computer Design: VLSI in Computers and Processors*, pp. 122-125.

[39] Neti, C., Schneider, M.H., and Young, E.D. (1992), "Maximally fault tolerant neural networks," *IEEE Transactions on Neural Networks*, Vol. 3, no. 1, pp. 14-23.

[40] Kim, J.H., Lursinsap, C., and Park, S. (1991), "Fault-tolerant artificial neural networks," *Int. Joint Conf. on Neural Networks (IJCNN'91)*, Vol. 2, pp. 951.

[41] Serrano-Gotarredona, T. and Linares-Barranco, B. (1997), "An ART1 microchip and its use in multi-ART1 systems," *IEEE Transactions on Neural Networks*, Vol. 8, no. 5, pp. 1184-1194.

[42] Carpenter, G.A., Grossberg, S., and Reynolds, J.H. (1991), "ARTMAP: Supervised real-time learning and classification of nonstationary data by a self-organizing neural network," *Neural Networks*, Vol. 4, pp. 759-771.

[43] Medeiro, F., Rodríguez-Macías, R., Fernández, F.V., Domínguez-Castro, R., and Rodríguez-Vázquez, A. (1994), "Global analogue cell design and optimization using statistical techniques," *Analog Integrated Circuits and Signal Processing*, Vol. 6, pp. 179-195.

[44] Bastos, J., Steyaert, M., Roovers, R., Kinger, P., Sansen, W., Graindourze, B., Pergoot, A., and Janssens, E. (1995), "Mismatch characterization of small size MOS transistors," *Proc. IEEE 1995 Int. Conf. Microelectronic Test Structures*, Vol. 8, pp. 271-276.

[45] Bastos, J., Steyaert, M., Pergoot, A., and Sansen, W. (1997), "Mismatch characterization of submicron MOS transistors," *Analog Integrated Circuits and Signal Processing*, Vol. 12, pp. 95-106.

[46] Rade, L. and Westergren, B. (1990), *BETA Mathematics Handbook*, CRC Press.

INDEX

Studies in Fuzziness and Soft Computing

Vol. 25. J. Buckley and Th. Feuring
Fuzzy and Neural: Interactions and Applications, 1999
ISBN 3-7908-1170-X

Vol. 26. A. Yazici and R. George
Fuzzy Database Modeling, 1999
ISBN 3-7908-1171-8

Vol. 27. M. Zaus
Crisp and Soft Computing with Hypercubical Calculus, 1999
ISBN 3-7908-1172-6

Vol. 28. R.A. Ribeiro, H.-J. Zimmermann, R.R. Yager and J. Kacprzyk (Eds.)
Soft Computing in Financial Engineering, 1999
ISBN 3-7908-1173-4

Vol. 29. H. Tanaka and P. Guo
Possibilistic Data Analysis for Operations Research, 1999
ISBN 3-7908-1183-1

Vol. 30. N. Kasabov and R. Kozma (Eds.)
Neuro-Fuzzy Techniques for Intelligent Information Systems, 1999
ISBN 3-7908-1187-4

Vol. 31. B. Kostek
Soft Computing in Acoustics, 1999
ISBN 3-7908-1190-4

Vol. 32. K. Hirota and T. Fukuda
Soft Computing in Mechatronics, 1999
ISBN 3-7908-1212-9

Vol. 33. L.A. Zadeh and J. Kacprzyk (Eds.)
Computing with Words in Information/ Intelligent Systems 1, 1999
ISBN 3-7908-1217-X

Vol. 34. L.A. Zadeh and J. Kacprzyk (Eds.)
Computing with Words in Information/ Intelligent Systems 2, 1999
ISBN 3-7908-1218-8

Vol. 35. K.T. Atanassov
Intuitionistic Fuzzy Sets, 1999
ISBN 3-7908-1228-5

Vol. 36. L.C. Jain (Ed.)
Innovative Teaching and Learning, 2000
ISBN 3-7908-1246-3

Vol. 37. R. Słowiński and M. Hapke (Eds.)
Scheduling Under Fuzziness, 2000
ISBN 3-7908-1249-8

Vol. 38. D. Ruan (Ed.)
Fuzzy Systems and Soft Computing in Nuclear Engineering, 2000
ISBN 3-7908-1251-X

Vol. 39. O. Pons, M.A. Vila and J. Kacprzyk (Eds.)
Knowledge Management in Fuzzy Databases, 2000
ISBN 3-7908-1255-2

Vol. 40. M. Grabisch, T. Murofushi and M. Sugeno (Eds.)
Fuzzy Measures and Integrals, 2000
ISBN 3-7908-1255-2

Vol. 41. P. Szczepaniak, P. Lisboa and J. Kacprzyk (Eds.)
Fuzzy Systems in Medicine, 2000
ISBN 3-7908-1263-4

Vol. 42. S.K. Pal, A. Ghosh and M.K. Kundu (Eds.)
Soft Computing for Image Processing, 2000
ISBN 3-7908-1268-4